Wastewater Pathogens

WASTEWATER MICROBIOLOGY SERIES

Editor

Michael H. Gerardi

Nitrification and Denitrification in the Activated Sludge Process
    Michael H. Gerardi

Settleability Problems and Loss of Solids in the Activated Sludge Process
    Michael H. Gerardi

The Microbiology of Anearobic Digesters
    Michael H. Gerardi

Wastewater Pathogens
    Michael H. Gerardi and Mel C. Zimmerman

Wastewater Pathogens

Michael H. Gerardi
The Pennsylvania State University

Mel C. Zimmerman
Lycoming College

WILEY-INTERSCIENCE

A John Wiley & Sons, Inc Publication

Library of Congress Cataloging-in-Publication Data:

Gerardi, Michael H.
    Wastewater pathogens / Michael H. Gerardi, Mel C. Zimmerman.
      p. cm.
    Includes bibliographical references and index.
    ISBN 0-471-20692-X (cloth)
    1. Waterborne infection. 2. Pathogenic microorganisms. 3. Sewage—Microbiology.
    4. Water—Microbiology. 5. Factory and trade waste—Health aspects. I. Zimmerman,
Melvin C. (Melvin Charles), 1950– II. Title.
    RA642.W3G47 2005
    363.72′84—dc22

2004015429

10 9 8 7 6 5 4 3 2 1

To
joVanna, Beth, and Elicia
Gail, Noah, and Drew

The authors extend their sincere appreciation to
joVanna Gerardi for computer support
and
Cristopher Noviello for artwork used in this text.

Contents

Preface

Because of the frequent, often daily, exposure to a large number and diversity of pathogens (disease-causing agents), wastewater personnel have general and specific concerns related to the potential health hazards presented by pathogens. Pathogens commonly found in wastewater and of concern to wastewater personnel include viruses, bacteria, fungi, protozoans, and helminths. Allergens, endotoxins, and exotoxins are also found in wastewater and represent a concern to wastewater personnel. Although most health hazards related to date with wastewater pathogens are minimal or nil, unique and potential hazards do exist, and much research work needs to be performed to better evaluate existing and potential hazards. Operator concerns, health hazards, and the emergence of new diseases such as bird flu virus, monkeypox, and West Nile virus provide the need for the development of a book reviewing wastewater pathogens.

This book is prepared for wastewater personnel who are exposed to wastewater, aerosols, sludge, compost, or contaminated surfaces and desire a practical review of wastewater pathogens, health hazards, and appropriate protective measures to guard against infection and disease. This book provides a review of wastewater pathogens, their sources and diseases, disease transmission, the fate of pathogens in wastewater collection and treatment systems, the body's defenses against infection, personal hygiene measures, protective equipment, and immunizations.

Wastewater Pathogens is the fourth book in the Wastewater Microbiology Series by John Wiley & Sons. This series is designed for wastewater personnel, and the series presents a microbiological review of the significant organisms and their roles in wastewater treatment facilities.

Michael H. Gerardi
Mel C. Zimmerman
Williamsport, Pennsylvania

Part I

Overview

1

Introduction

Wastewater collection and treatment have greatly reduced the number of outbreaks of disease in the United States each year. Wastewater treatment consists of a combination of biological, chemical, and physical processes to degrade organic and nitrogenous wastes and destroy or inactivate a large number of pathogens. Usually, chlorine (Cl_2) or chlorinated compounds, ozone (O_3), or ultraviolet light (uv) is used to disinfect the treated effluent before its discharge to receiving waters.

Every stream and lake has some limited capacity to degrade wastes and reduce the number of pathogens by natural inactivation and destruction processes. Inactivation or destruction is achieved through adsorption, predation, dilution, change in water temperature, and solar radiation. Because of the large quantity of effluent discharged to the receiving waters, the natural processes of pathogen reduction are inadequate for protection of public health. In addition, industrial wastes that alter the water pH and provide excessive bacterial nutrients often compromise the ability of natural processes to inactive and destroy pathogens. Therefore, the disinfection of effluent has assumed critical importance.

The degradation of organic and nitrogenous wastes by biological wastewater treatment plants (activated sludge and trickling filter) results in the production of large quantities of sludge. Numerous pathogens are contained in the sludge. Many of the pathogens are inactivated or destroyed by additional biological, chemical, and physical processes before its disposal.

Pathogens in wastewater and sludge represent health hazards to individuals working at wastewater treatment facilities and sludge disposal sites. The pathogens also represent health hazards to community members living downwind of wastewater treatment facilities and near sludge disposal sites.

A large number and diversity of pathogens are found in wastewater and sludge. The pathogens include viruses, bacteria, fungi, protozoans, helminths (worms),

Wastewater Pathogens, by Michael H. Gerardi and Mel C. Zimmerman
ISBN 0-471-20692-X Copyright © 2005 John Wiley & Sons, Inc.

allergens, and toxins. The most prevalent are viruses, bacteria, fungi, protozoans, and helminths. The pathogens are a reflection of the common diseases in a community and may be present in numbers as great as 100,000,000 per milliliter. The pathogens of greatest concern to wastewater personnel are enteric viruses, enteric bacteria, especially *Campylobacter*, the bacterium *Leptospira*, the fungus *Aspergillus*, the protozoans *Giardia* and *Cryptosporidium*, and the tapeworm *Hymenolepis*.

Pathogens enter wastewater treatment facilities from several sources. Pathogenic agents are found in fecal waste and urine from humans and animals. They can enter wastewater treatment facilities from humans who are sick or are carriers of disease. Wastewater from baths, dishwashers, showers, sinks, and washing machines also may contain pathogens. Viruses in the feces of an infected host may be present at concentrations as high as 10,000,000,000 per gram of wet weight of feces. Pathogenic bacteria in the feces of an infected host may be present at concentrations as high as 1,000,000 per gram of wet weight of feces.

Pathogens in animal waste on the ground can enter wastewater treatment facilities through inflow and infiltration (I/I). Animal wastes from meatpacking and processing facilities and from rats in sanitary sewers also serve as sources of pathogens.

Waterborne, foodborne, bloodborne [hepatitis B virus and human immunodeficiency virus (HIV)], and sexually transmitted pathogens are found in wastewater. Along with these pathogens, microbial toxins and allergens also are found. Viruses (Table 1.1) commonly found in wastewater include hepatitis and the human immunodeficiency virus.

Pathogenic bacteria (Table 1.2) commonly found in wastewater include *Leptospira* and *Salmonella*. Pathogenic fungi (Table 1.3) commonly found in wastewater include *Aspergillus* and *Candida*.

Disease-causing or parasitic protozoans (Table 1.4) commonly found in wastewater include *Giardia lamblia* and *Cryptosporidium parvum*. Parasitic helminths (Table 1.5) that have been found in relatively large numbers in wastewater include cestodes (tapeworms) such as *Taenia* and nematodes (roundworms) such as *Ascaris* and *Trichuris*. The U.S. Environmental Protection Agency (U.S. EPA) and Centers

TABLE 1.1 Viruses/Viral Groups Commonly Found in Wastewater

Virus/Viral Group	Disease
Adenovirus	Upper and lower respiratory tract distress
Coxsackievirus	Common cold and pharyngitis
Enterovirus	Upper respiratory tract distress and gastroenteritis
Hepatitis	Hepatitis
HIV	Acquired immunodeficiency syndrome (AIDS)
Influenza	"Flu"
Norwalk	Gastroenteritis
Poliovirus	Poliomyelitis
Reovirus	Upper respiratory tract distress and gastroenteritis
Rotavirus	Gastroenteritis

TABLE 1.2 Pathogenic Bacteria Commonly Found in Wastewater

Bacterium/Bacteria	Disease
Actinomyces israelii	Actinomyocsis
Bacillus anthracis	Anthrax
Brucella spp.	Brucellosis (Malta fever)
Campylobacter jejuni	Gastroenteritis
Clostridium spp.	Gas gangrene
Enterotoxigenic *Escherichia coli* (ETEC)	Gastroenteritis, diarrhea
Francisella tularensis	Tularemia
Leptospira interrogans icterohemorrhagiae	Leptospirosis (Weil disease)
Mycobacterium tuberculosis	Tuberculosis
Nocardia spp.	Nocardiosis
Salmonella enterica paratyphi	Paratyphoid fever
Salmonella spp.	Salmonellosis (food poisoning)
Salmonella typhi	Typhoid fever
Shigella spp.	Shigellosis (bacillary dysentery)
Vibrio cholerae	Cholera (Asiatic cholera)
Vibrio parahaemolyticus	Gastroenteritis
Yersinia enterocolitica	Yersiniosis (gastroenteritis)

TABLE 1.3 Pathogenic Fungi Commonly Found in Wastewater

Fungus/Fungi	Disease
Aspergillus fumigatus	"Farmer's lung" (allergic lung disease)
Candida spp.	Candidiasis

TABLE 1.4 Parasitic Protozoans Commonly Found in Wastewater

Protozoan	Disease
Balantidium coli	Balantidosis
Cryptosporidium parvum	Cryptosporidiosis (diarrhea)
Entamoeba coli	Diarrhea, ulceration
Entamoeba histolytica	Amebiasis (amebic dysentry)
Giardia lamblia	Giardiasis (diarrhea, malabsorption)

TABLE 1.5 Parasitic Helminths Commonly Found in Wastewater

Helminth	Type of Worm	Disease
Taenia saginata	Tapeworm	Taeniasis
Ancylostoma spp.	Roundworm	Anemia
Ascaris spp.	Roundworm	Ascariasis
Echinococcus granulosus	Tapeworm	Echinococcosis
Enterobius spp.	Roundworm	Enterobiasis
Necator americanus	Roundworm	Hookworm disease (anemia)
Schistomsoma spp.	Flatworm	Schistosomiasis (swimmer's itch)
Strongyloides stercoralis	Roundworm	Strongyloidiasis
Taenia spp.	Tapeworm	Taeniasis
Trichuris spp.	Roundworm	Anemia, diarrhea

for Disease Control and Prevention (CDC) maintain data on many of these proto-zoan and helminthic diseases as well as viral, bacterial, and fungal diseases. This information as well as data on other diseases also is maintained by CDC and state and local health departments.

2

History

Over the last 150 years wastewater collection and treatment systems have been designed and built for three purposes: (1) to provide clean water for cities undergoing rapid industrialization, (2) to protect the quality of the waters receiving the effluent from the treatment plant, and (3) to control the outbreaks of communicable diseases. Outbreaks of communicable diseases often were related to poor sanitary conditions.

Before wastewater treatment was required, raw wastewater was discharged directly into steams and lakes by "wildcat" sewers. Despite significant and prolonged environment damage, major environment legislation did not take effect until the 1970s.

Initially, channels were used to collect and convey wastewaters and mechanical processes (sedimentation) and chemicals (flocculants) were used to remove wastes from the wastewater. Although the collection and conveyance of wastewater and the removal of wastes improved the living conditions of cities, these measures often moved wastes from cities to bodies of water more rapidly than the receiving water could treat the wastes. Also, the exposure of wastewater personnel to the large number and diversity of pathogenic agents in wastewater was exacerbated by expanding human populations and the trend toward concentrated domestic animal husbandry.

In 1972 the Federal Water Pollution Act was passed. This act required communities to treat wastewater in order to protect human health and environmental quality. As a result of this legislation, more efficient measures for treating wastes became necessary.

The need for more efficient treatment of wastes was satisfied with the development of biological treatment processes. The trickling filter process (Fig. 2.1) was first

Wastewater Pathogens, by Michael H. Gerardi and Mel C. Zimmerman
ISBN 0-471-20692-X   Copyright © 2005 John Wiley & Sons, Inc.

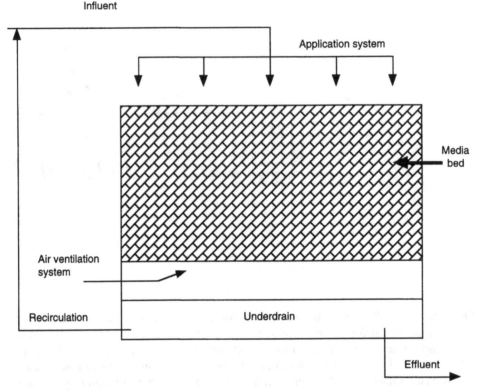

Influent

Application system

Media bed

Air ventilation system

Recirculation

Underdrain

Effluent

Figure 2.1   *Trickling filter process. The trickling filter process consists of a media bed, usually made of rock or plastic, that supports a film of biological growth (biofilm), mostly bacterial. Influent wastewater is applied to the surface of the media bed. Once applied, the wastewater percolates over the surface of the biological growth. The bacteria in the biofilm degrade carbonaceous and nitrogenous wastes, and particulate and colloidal wastes as well as heavy metals also are removed. The effluent from the trickling filter usually undergoes further treatment such as disinfection or may be recirculated to the surface of the media bed for additional treatment. In domestic and municipal trickling filter processes, the influent, effluent, and biofilm contain pathogens.*

used to treat wastewater. This process eventually was replaced with the activated sludge process (Fig. 2.2). Since the 1970s, numerous activated sludge processes have been put in operation.

Wastes degraded by these biological treatment processes resulted in the growth of bacteria (sludge production). Wastes collected but not degraded by the activated sludge and trickling filter processes were degraded with the use of additional biological units, aerobic and anaerobic digesters. The additional treatment resulted in the production of more sludge. Sludge produced by most biological wastewater treatment processes required disposal. Sludge disposal options are affected by several factors including impact on environmental quality and human health.

Treatment of wastes to render them nonpathogenic became necessary. Wastewater treatment systems were developed in part to control the outbreak of communicable diseases. Wastewater collection became wastewater treatment and sludge

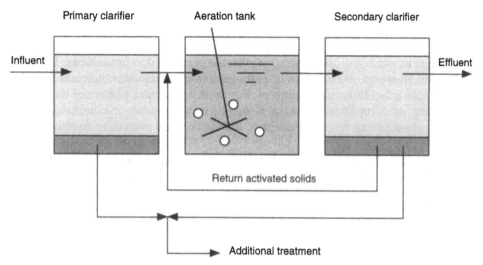

Primary clarifier   Aeration tank   Secondary clarifier

Influent

Effluent

Return activated solids

Additional treatment

Figure 2.2 *Activated sludge process. A typical activated sludge process consists of an aeration tank where influent wastewater is mixed with oxygen and bacteria. The bacteria are present in the aeration tank in flocculated masses or floc particles. The bacteria in the floc particles degrade carbonaceous and nitrogenous wastes, and particulate and colloidal wastes as well as heavy metals also are removed. Downstream of the aeration tank is a secondary sedimentation tank or secondary clarifier. In the quiescent environment of the clarifier the floc particles settle to the bottom of the clarifier. The liquid or supernatant above the settled solids usually undergoes further treatment such as disinfection. The settled solids may be returned to the aeration tank or wasted from the system for additional treatment and disposal. Many activated sludge processes have a sedimentation tank or clarifier upstream of the aeration tank. In the quiescent environment of the primary clarifier many solids in the wastewater settle to the bottom of the clarifier. Solids removed from the primary clarifier undergo further treatment and eventual disposal. In domestic and municipal activated sludge processes, the solids from the primary clarifier and secondary clarifier as well as the activated sludge and secondary clarifier effluent contain pathogens.*

disposal, and new systems were used for the treatment of wastewater and disposal of sludge. Although these systems control the outbreaks of disease, they do not eliminate all the pathogens responsible for disease. The systems concentrate the numbers of pathogens. The concentration of pathogens by biological treatment systems represents a potential health hazard to wastewater personnel.

With the recognition of the role of pathogens in causing disease, disinfection of the treated wastewater or effluent was instituted. Today, chlorination of effluent is used at many wastewater treatment plants. However, chlorination results in the production of undesirable chemical compounds. Because of the production of these compounds, chlorination of the effluent gradually is being replaced by other disinfection techniques such as ultraviolet (uv) irradiation.

All of the processes used for the collection and treatment of wastewaters concentrate and produce an incomplete elimination of pathogens. These processes leave it to the "natural" cleansing properties of the receiving water and soil (sludge-amended land) to further reduce the number of pathogens. The collection and concentration of pathogens and their incomplete elimination expose wastewater personnel to the risk of disease.

Because of their daily exposure to wastewater, aerosols, sludge, and contaminated surfaces, wastewater personnel have a higher incidence of exposure to pathogens than the general public. The risk to wastewater personnel of becoming infected with a pathogen at work sites and contracting disease is minimal to insignificant for most pathogens. However, the isolation of pathogens from wastewater indicates that the potential for a health hazard is present. For most wastewater personnel the risk of developing an occupational disease is significantly reduced or eliminated when common sense, proper hygiene measures, and appropriate protective equipment are used.

To better assess risk for wastewater personnel, etiologic studies are needed. The studies should be designed to isolate a pathogen from an affected individual, such as the bacterium responsible for typhoid fever (*Salmonella typhi*), or demonstrate the production of specific antibodies in an affected individual, such as those for the bacterium responsible for leptospirosis (*Leptospira interrogans*). Also, there is a need to examine the long-term effects of exposure of wastewater personnel to pathogens in wastewater.

Although wastewater personnel have a relatively low risk of contracting many of the diseases associated with pathogens in wastewater, occupational disease can occur without symptoms. The common symptoms of infection among wastewater personnel include eye and nose irritation, lower respiratory tract problems, fever, fatigue, skin irritation, headaches, dizziness, and flulike conditions. Whether or not wastewater personnel will become ill after being exposed to wastewater is hard to predict. However, there are enough pathogens in wastewater to make exposure to wastewater risky, that is, the risk of infection is real. Therefore, complacency with respect to exposure to wastewater and failure to use proper hygiene measures and appropriate protective equipment can be dangerous.

3

Hazards

Wastewater personnel are exposed to numerous occupational hazards. These hazards include gases, endotoxins and exotoxins, chemicals, allergens, and pathogens.

GASES

The degradation of wastewater and sludge results in the release of several gases (Table 3.1) that are asphyxiating, irritating, toxic, or flammable. These gases include ammonia (NH_3), carbon dioxide (CO_2), carbon monoxide (CO), hydrogen sulfide (H_2S), and methane (CH_4).

Ammonia is a strong respiratory tract irritant. Because ammonia is highly soluble, it dissolves in the water film covering the mucous membranes and the upper respiratory tract. At high concentrations (>100 ppm) ammonia can irritate the lower respiratory tract and cause respiratory distress. Ammonia is lighter than air and collects in the upper levels of confined spaces.

Carbon dioxide is an asphyxiant. When carbon dioxide accumulates to more than 10% of the atmosphere, narcosis may occur in individuals breathing the carbon dioxide. Carbon dioxide is heavier than air and collects in the lower levels of confined spaces.

Carbon monoxide also is an asphyxiant. Carbon monoxide quickly bonds to hemoglobin, resulting in a decrease in the amount of oxygen carried throughout the body. With a decrease in oxygen transportation throughout the body, headaches develop. Increased exposure to carbon monoxide may result in collapse, coma, and death. Carbon monoxide also is heavier than air and collects in the lower levels of confined spaces.

Wastewater Pathogens, by Michael H. Gerardi and Mel C. Zimmerman
ISBN 0-471-20692-X   Copyright © 2005 John Wiley & Sons, Inc.

TABLE 3.1 Hazardous Gases Produced Through Microbial Activity at Wastewater Treatment Systems

Gas	Formula	Lighter/Heavier than Air	Significance
Ammonia	NH_3	Lighter	Irritant; causes respiratory distress
Carbon dioxide	CO_2	Heavier	Asphyxiant; causes narcosis
Carbon monoxide	CO	Heavier	Asphyxiant; causes oxygen deficiency
Hydrogen sulfide	H_2S	Heavier	Irritant and asphyxiant; inhibits respiratory tract
Methane	CH_4	Lighter	Asphyxiant and flammable

TABLE 3.2 Examples of Volatile Organic Compounds

Compound	Formula
Cadaverine	$H_2N(CH_2)NH_2$
Ethyl mercaptan	C_2H_5SH
Formaldehyde	CH_2O
Formic acid	HCOOH
Indole	$C_8H_{13}N$
Methyl mercaptan	CH_3SH
Putresine	$H_2N(CH_2)_4NH_2$
Skatole	C_9H_9N
Trimethylamine	$CH_3NCH_3CH_3$

Hydrogen sulfide is a strong respiratory tract irritant and asphyxiant. Hydrogen sulfide has a characteristic "rotten egg" odor and is highly soluble in water. Because of its high solubility, hydrogen sulfide irritates the mucous membranes of the respiratory tract and the eyes.

A high concentration of hydrogen sulfide cause olfactory paralysis, that is, an exposed individual loses the ability to smell hydrogen sulfide. Prolonged exposure to hydrogen sulfide results in a loss of consciousness. At concentrations >100 ppm, hydrogen sulfide inhibits the respiratory tract and causes death. Hydrogen sulfide is heavier than air and collects in the lower levels of confined spaces.

Methane is produced from the anaerobic degradation of wastewater and sludge and is colorless, odorless, and highly flammable. Methane is an asphyxiant and is lighter than air. The gas collects in the upper levels of confined spaces.

Vaporized volatile organic compounds (VOCs) also represent a health risk (Table 3.2). These compounds are produced through the anaerobic degradation of wastes, particularly nitrogen-containing and sulfur-containing proteins.

BIOLOGICAL TOXINS

Numerous Gram-negative and Gram-positive bacteria are found in wastewater and sludge and as aerosolized organisms. Toxins produced by these bacteria are known

TABLE 3.3   Release of Endotoxins and Endotoxins

Toxin	Released by Dead/Living Cell	Bacterium
Endotoxin	Dead	Nearly all Gram negative
Exotoxin	Living	Mostly Gram positive

TABLE 3.4   Comparison of Major Characteristics of Endotoxins and Exotoxins

Characteristic	Endotoxins	Exotoxins
Bacteria producing toxin	Nearly all Gram negative	Mostly Gram positive
Location of toxin	Released from cell wall of dead bacteria	Released by living bacteria
Major chemical component	Lipopolysaccharide	Protein
Reaction in individuals	Nonspecific and localized	Specific
Toxicity	Usually weakly toxic	Strongly toxic
Example of disease	Salmonellosis	Tetanus

TABLE 3.5   Examples of Bacteria That Produce Exotoxins

Bacterium	Disease
Bacillus anthracis	Anthrax
Bacillus cereus	Enterotoxicosis (food poisoning)
Clostridium tetani	Tetanus (lockjaw)
Corynebacterium diphtheriae	Diphtheria
Escherichia coli	Enterotoxicosis
Staphylococcus aureus	Enterotoxicosis
Vibrio cholerae	Cholera

as endotoxins and exotoxins (Table 3.3). Many of these toxins cause gastrointestinal or respiratory tract diseases.

Bacteria release endotoxins at the time of their death and autolysis. These toxins (Table 3.4) are components of cell walls. As the bacterium dies, lysis occurs, that is, the cell wall deteriorates and the toxins disperse into the host's tissues. Endotoxins are heat-stable lipopolysaccharides that cause nonspecific or localized reactions in individuals.

Living bacteria release exotoxins (Table 3.4). The toxins are excreted by the bacteria into their surrounding medium and are absorbed by the host's tissues. Exotoxins are proteinaceous molecules. Although they are heat stable, they are not as stable as endotoxins. Exotoxins are very potent toxins and highly specific with respect to the reactions they cause in individuals, for example, they attack the nervous system (neurotoxins) and heart muscles (cardiac muscle toxins). Examples of diseases caused by exotoxins include botulism, staphylococcal food poisoning, and tetanus (Tables 3.5 and 3.6).

Exotoxins that attack the intestinal tract are known as enterotoxins. These proteinaceous molecules are toxin specific for the cells of the intestinal mucosa. Certain

TABLE 3.6   Examples of Bacteria or Genera of Bacteria that Cause Food Poisoning

Bacterium/Bacterial Genus	How Acquired (raw or improperly stored or prepared foods)
Campylobacter	Poultry
Escherichia coli 0157:H7	Hamburger, alfalfa sprouts, unpasteurized fruit juices, dry-cured salami, lettuce
Listeria	Soft cheese, unpasteurized milk, imported seafood products, frozen cooked crab meat, cooked shrimp
Shigella	Milk and dairy products
Vibrio	Seafood
Yersinia	Pork

TABLE 3.7   Common Allergens

Portal of Entry to the Body	Examples
Ingestion	Eggs, fruits, medication (aspirin, penicillin, sulfur-containing drugs) milk, nuts, peanut butter, seafood
Inhalation	Dander, mites and their fecal pellets, pesticides, pollen (grasses, trees, weeds), spores (bacterial, fungal)
Invasion (injection)	Antibodies, hormones, venoms (insects, snakes, spiders)

types of *Staphylococcus aureus* cause food poisoning (enterotoxicosis). Because these bacteria are heat stable and resistant to desiccation, foods easily become contaminated with *Staphylococcus aureus*.

CHEMICALS

Wastewater treatment personnel are exposed to a large variety of chemicals. These chemicals include those used at the treatment system such as chlorine (Cl_2) and sulfur dioxide (SO_2) and those found in the wastewater. These compounds include cleaners, solvents, lubricants, caustics, acids, and pesticides.

ALLERGENS

An allergen is any ordinary innocuous foreign substance that can elicit an adverse immunologic response in a sensitized person (Table 3.7). Allergens may elicit a response through ingestion, inhalation, or invasion. Examples of allergens include antibiotics, dander, dust, feathers, hair, mites and their fecal pellets, pollen, and certain foods.

PATHOGENS

Although wastewater and wastewater treatment processes are hostile environments for pathogens, many viruses and pathogenic organisms survive these environments.

The surviving pathogens usually are more resistant to these hostile environments than many other microscopic organisms.

Airborne, bloodborne, foodborne, and waterborne pathogens can be present in wastewater and wastewater treatment processes. Pathogens consist of a variety of viruses, bacteria, fungi, helminths, and protozoans.

4

Classification of Organisms

The large diversity of organisms is organized or classified into many categories according to the characteristics they share with other organisms. These characteristics are similarities in such factors as structure, genetics, biochemistry, and reproduction. The major categories that are use to classify organisms consist of the following:

- Kingdom
- Phylum
- Class
- Order
- Family
- Genus
- Species

Collectively, these categories make up a hierarchy of categories with the first or top category, the Kingdom, being a very broad, general category. With each descending or lower category, the number of organisms that match the characteristics for that category decreases. At the last or bottom category, the Species, only one organism fits the characteristics. For example, there are several organisms of concern or interest to wastewater personnel that are pathogens or carriers (vectors) of pathogens that are reviewed in this text and classified in Table 4.1. These organisms are the protozoan *Giardia lamblia*, the tapeworm *Hymenolepis nana*, the mosquito *Culex pipiens* (vector for West Nile virus), and the sewer rat *Rattus norvegicus* (vector for numerous pathogens). Viruses can be pathogens but they can be not

Wastewater Pathogens, by Michael H. Gerardi and Mel C. Zimmerman
ISBN 0-471-20692-X   Copyright © 2005 John Wiley & Sons, Inc.

TABLE 4.1   Classification of Some Pathogenic Agents and Vectors of Pathogenic Agents

Category	Organism			
	Giardia lamblia	*Hymenolepis nana*	*Culex pipiens*	*Rattus norvegicus*
Kingdom	Protista	Animal	Animal	Animal
Phylum	Sarcomastigophora	Platyhelminthes	Arthropoda	Chordata
Class	Zoomastigophora	Cestoidea	Insecta	Mammalia
Order	Diplomonadida	Cyclophyllidea	Diptera	Rodentia
Family	Hexamitidae	Hymenolepidae	Culcidae	Muridae
Genus	*Giardia*	*Hymenolepis*	*Culex*	*Rattus*
Species	*lamblia*	*nana*	*pipiens*	*norvegicus*

organisms. Characteristics used for classifying viruses are presented in Chapter 5, *Viruses*. Bacteria are organisms, but they are classified differently than other organisms. The classification of bacteria is presented in Chapter 9, *Bacteria*.

When reference is made to a specific organism, for example, *Giardia lamblia*, the complete classification hierarchy is not used. The organism is identified by its scientific name or bionomen. This name consists of the generic name (genus category) and specific name (species category). The scientific name distinguishes the organisms from other species in the genus. Examples of scientific names are *Giardia lamblia*, *Hymenolepis nana*, *Culex pipiens*, and *Rattus norvegicus*.

When a scientific name is used, the genus name is given first and the species name second. The genus name always is capitalized. The scientific name or bionomen is placed in bold or italicized type or underlined. If the scientific name is repeated, the genus name may be abbreviated using the first letter of the genus name, for example, *G. lamblia*, *H. nana*, *C. pipiens*, and *R. norvegicus*. When the name or names of the species are not known or provided, the abbreviations "sp." and "spp." are used to reference one species or two or more species, respectively.

As more and more organisms are identified and improved techniques are developed to better categorize organisms, more elaborated classification schemes or hierarchies are used. New hierarchies consist of the following categories:

- Kingdom
- Phylum
- Subphylum
- Superclass
- Class
- Subclass
- Cohort
- Superorder
- Order
- Superfamily
- Family
- Subfamily

- Tribe
- Genus
- Subgenus
- Species
- Subspecies

Part II

Viruses, Bacteria, and Fungi

5

Viruses

Viruses are ultramicroscopic agents. They cannot be observed with a conventional microscope. They must be observed with an electron microscope. Viruses are inert or nonliving and lack mobility.

Viruses are diverse in structure. There are two basic physical components that make up all viruses. These components are genetic material (core) and a protein coat or capsid (cover). The coat provides a protective layer for the virus and "recognizes" the correct host cell to be attacked. When the genetic material of the virus is introduced into a host cell, the genetic material takes control of the reproductive mechanism of the cell and causes the cell to produce viruses, not cells. In addition to these two basic components, some viruses have an additional protective layer, the lipid envelope.

A major strength of viruses is their ability to mutate (change) quickly. Mutation often provides improved protection from harsh environments and vaccines. A frequent mutation of viruses is their ability to change their proteinaceous coat. Retroviruses, including the human immunodeficiency virus (HIV) readily undergo changes in their coat. This is a major factor in the development of viral resistance to vaccines and antiviral drugs.

The genetic material of viruses consists of either ribonucleic acid (RNA) or deoxyribonucleic acid (DNA) (Fig. 5.1). The genetic material makes up the core of the virus. The capsid surrounds and protects the genetic material. Each capsid has its own unique combination of proteins. A major weakness of viruses is that they have no independent ability to repair DNA or RNA damage. Therefore, ionization (ionizing radiation) can easily destroy viruses.

Viruses are not capable of independent growth or reproduction, and therefore they are not living organisms. Viruses increase in number through replication. For replication to occur, the virus must first attach to or enter a living host cell, for

Wastewater Pathogens, by Michael H. Gerardi and Mel C. Zimmerman
ISBN 0-471-20692-X   Copyright © 2005 John Wiley & Sons, Inc.

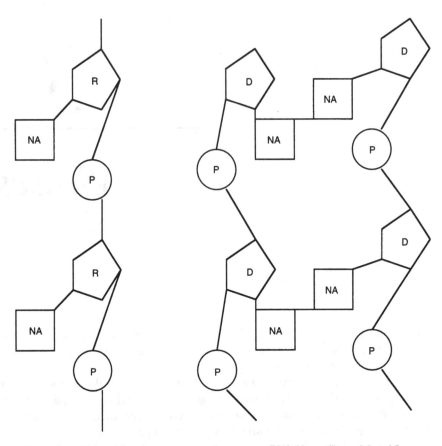

RNA (ribonucleic acid)   DNA (deoxyribonucleic acid)

Components of RNA and DNA

Component	RNA	DNA
Phosphoric acid	X	X
Sugar - ribose	X	
Sugar - deoxyribose		X
Base - adenine	X	X
Base - cytosine	X	X
Base - guanine	X	X
Base - thymine		X
Base - uracil	X	X

Figure 5.1   *RNA and DNA. RNA and DNA represent the genetic information of viruses and cells. However, there are significant differences between RNA and DNA. RNA is a single strand of "information," whereas DNA is a double strand of "information." RNA and DNA each have a 5-carbon sugar, but the sugar in RNA (ribose or "R") and the sugar in DNA (deoxyribose or "D") are different in structure. A phosphate ("P") group couples the sugar groups in each strand of information. Although each strand of information contains nucleic acids ("NA") or bases, there are some differences in the bases of each strand. Most importantly, the bases in the DNA strand of information are paired.*

example, a bacterial or human cell, and then must transfer its genetic material (RNA or DNA) to the cell. Because viruses cannot replicate outside of a host, detection of viruses depends on cell infection assays or molecular techniques for the presence of viral DNA or RNA.

Viruses attach to a host cell by the capsid and transfer their genetic material to the cell by injecting their RNA or DNA. Viruses can bind and enter only specific cells, for example, liver cells in humans. Different cells present different external structures that viruses must penetrate to cause infection. The cell membrane is the structural unit that viruses must penetrate in all organisms. Although genetic material enters the host cell of humans, animals, and protozoa through attachment, viruses have additional mechanisms of entering host cells that have a cell wall, for example, bacterial cells. The virus may penetrate or enter these cells by fusion with the cell or by channel formation through the cell wall.

Once injected, the viral RNA or DNA "takes over" the reproductive machinery of the host cell. The injection of viral RNA or DNA results in an infection of the cell. Once infected, the host cell produces or replicates numerous copies of the virus. When the replicated copies of the virus leave the host cell, the cell is damaged or destroyed. The damage to or destruction of the cell results in disease. The released viruses continue to infect new cells and spread the disease.

Because of their means of replication, viruses are considered to be obligate intracellular parasites, that is, all viruses are pathogens. Viruses infect all organisms. Although all viruses are parasitic, some viruses have beneficial value. Some are useful sources of antibiotics.

Viruses that attack bacteria are bacteriophages, viral parasites of pathogenic bacteria. Bacteriophages are specific for a particular genus or species of bacteria. Some bacteriophages are being studied as a treatment technique for antibiotic-resistant bacteria such as *Staphylococcus aureus*.

Viruses are classified by several characteristics (Table 5.1). Major characteristics include type of genetic material (RNA or DNA), protein composition of the capsid, and presence or absence of an envelope. Additional characteristics that are used to classify viruses include the lipid composition of the envelope, shape and size of the virus, host affinities, and tissue or cell tropism.

Most viruses are helical (tubular), icosahedral (quasi-spherical), or complex in shape. Viruses are too small to be seen with a conventional or bright-field microscope. Therefore, the image or shadow of a virus is observed with an electron microscope.

The size of a virus is measured in nanometers (nm) (Fig. 5.2). A nanometer equals 0.001 microns (μm). A large virus such as the smallpox virus is nearly 1 μm in size.

TABLE 5.1 Characteristics Used to Classify Viruses

Genetic material
Protein coat or capsid
Presence or absence of an envelope
Lipids in the envelope
Shape of the virus
Size of the virus
Host affinities
Tissue or cell tropism

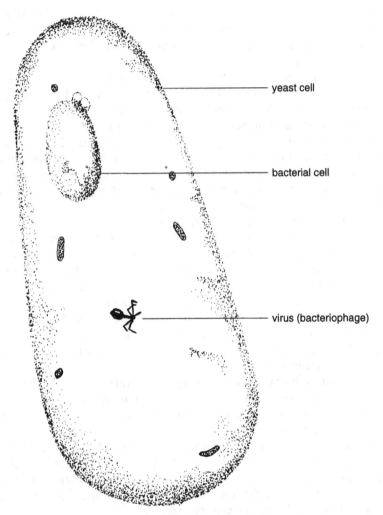

Figure 5.2 *Size of viruses (drawing not to scale). Compared with cells such as yeast cells and bacterial cells, viruses are very small. A virus would appear as a small dot on a bacterial cell and an even smaller dot on a yeast cell. Yeast cells and bacterial cells are measured in microns (μm), whereas viruses are measured in nanometers (nm). A nanometer is 1,000th of a micron.*

This large virus is approximately the same size as the bacterium *Escherichia coli* that is commonly found in wastewater and the intestinal tract of humans and animals. A small virus such as the yellow fever virus is approximately 10 nm (0.01 μm) in size. The yellow fever virus is similar in size to a small protein molecule. Most viruses range in size from 20 to 300 nm.

Host affinities of viruses are specific. Viruses attack a specific host or group of hosts. For example, some viruses only attack bacteria. These viruses are known as bacteriophages. Some viruses, such as the tobacco mosaic virus, only attack specific plants. And, unfortunately, some viruses attack humans.

Viruses are not only host specific but also tissue or cell specific. For example, the hepatitis viruses attack only the liver, the human immunodeficiency virus (HIV)

TABLE 5.2 Major Human Viral Groups of Concern to Wastewater Personnel

Viral Group	Disease
Adenovirus	Respiratory distress
Enterovirus: coxsackievirus A	Muscle distress
Enterovirus: coxsackievirus B	Muscle distress
Enterovirus: echovirus	Common cold
Enterovirus: poliovirus	Poliomyelitis
Hepatitis viral group (A, B, C, D, E, and F)	Hepatitis
Influenza virus	Influenza
Reovirus	Respiratory distress and gastroenteritis
Rotavirus	Enteritis

TABLE 5.3 Modes of Transmission of Several Viruses

Virus Transmission	Mode
Adenovirus (respiratory)	Inhalation
Echo (respiratory)	Inhalation
Enterovirus	Ingestion (enteric)
Hepatitis A	Ingestion (enteric)
Poliomyelitis	Ingestion (enteric)
Rotavirus	Ingestion (enteric)
West Nile virus (mechanical)	Invasion

attacks the cells of the immune system, and the poliovirus attacks the nervous system.

Viruses serve no useful purpose in wastewater treatment processes. They do not participate in the removal of wastes. Viruses also are not normal inhabitants of the human intestinal tract. However, in individuals who are infected with viruses, the concentration of viruses in the feces may range from 10^6 to 10^{10} viruses per gram of wet feces.

The concentration of viruses in wastewaters in the United States varies widely. However, the concentration of viruses increases with seasonal increases in enteric viral infections.

There are several major human viral groups of concern to wastewater personnel (Table 5.2). Of these groups, the enteric viruses (enteroviruses), including the hepatitis viruses, are of most concern. There are over 100 enteric viruses that are pathogenic to humans at a relatively low infectious dose. Significant enteric viruses include coxsackieviruses and echoviruses. In addition to these groups, there are several emerging viruses of interest. These viruses include the bird flu virus, HIV, and the West Nile virus (WNV).

Viruses may be transmitted by several means (Tables 5.3 and 5.4). Transmission often occurs through the intestinal mucosa (fecal-oral route or enteric) or the respiratory mucosa (direct inhalation of aerosols or respiratory). Transmission through the intestinal mucosa is the more common route. Mechanical vectors such as flies

TABLE 5.4   Examples of Viral Disease and Their Modes of Transmission

Disease	Virus	Mode of Transmission
AIDS	HIV	Sexual contact; mother-to-child; blood products; sharing needles; health care injuries
Chickenpox	Varicella-zoster virus	Airborne droplets; direct contact
Common cold	Rhinovirus	Airborne droplets; hand-to-hand contact
Hepatitis A	Hepatitis A	Contaminate food or water
Hepatitis B	Hepatitis B	Sexual contact; bloodborne
Influenza	Influenza A, B, C	Airborne droplets
Mononucleosis	Epstein–Barr virus	Saliva
Rabies	Arhabdovirus	Bite by infected animal

TABLE 5.5   Emerging Viral Diseases

Disease	Animal Vector
Bird flu	Close contact with infected poultry
Lyme disease	Deer tick
Monkeypox	Close contact with (getting bitten by) infected animals, especially prairie dogs
SARS	Unknown; infected animals suspected
West Nile virus	Bites from mosquitoes that have bitten infected birds, especially crows and jays

and mosquitoes also may transmit viruses. The concept of transmission through mechanical vectors also draws attention to the need for proper housekeeping and proper personal hygiene.

Wastewater personnel have a high incidence of exposure to enteric viruses. The enteric viruses replicate in the intestinal tract of humans and are shed in the fecal waste of infected individuals. These viruses are shed in the feces of infected individuals for several weeks.

Antibiotics are ineffective against viruses. Although several antiviral medications are available, the best defenses against viruses are the use of common sense, proper personal hygiene measures, and appropriate protective equipment and immunization. Immunization for many viruses can be obtained through several means including vaccination.

There are numerous concerns related to the hepatitis virus group, HIV, and several emerging viral diseases that are transmitted by animal vectors and include monkeypox, the bird flu virus, and West Nile virus (Table 5.5). Of the concerns related to these diseases, the possibility of epidemics is significant.

Factors that favor epidemics of these emerging diseases include international trade in exotic animals, global travel, and the loss of natural animal habitats to agriculture and development. For an epidemic to occur, the virus-infected animal must come in contact with humans. For example, in the southwest United States a hantavirus outbreak occurred in 1993. A good crop season in 1993 produced a bumper yield of seeds that led to rapid and large population explosions of deer mice (*Peromyscus maniculatus*). The deer mouse carried and excreted the virus, and the virus was then carried in dust. Humans came in contact with the virus-contaminated dust when the deer mouse built nests in homes.

MONKEYPOX

Monkeypox is related to smallpox and causes a nearly identical disease. Monkeypox was first reported in the 1990s in the Democratic Republic of Congo. It is believed that monkeypox was brought into the United States by exotic animals and is transferred from animals such as prairie dogs to humans. Humans, primates, rodents, and rabbits are most susceptible to monkeypox.

Monkeypox causes fever, cough, swollen lymph nodes, and lesions. Although monkeypox is related to smallpox, it is not as contagious or as deadly as smallpox. The smallpox vaccine protects against monkeypox.

BIRD FLU VIRUS

Influenza causes many American deaths each year. Bird flu (influenza) is one of the latest emerging diseases. Bird flu is known also as Asian flu or avian flu. Bird flu is caused by one of 15 avian flu viruses. The virus is worrisome for several reasons. It mutates rapidly and obtains genetic information from other flu viruses that infect other animals and humans. The virus is dangerous to humans and spreads quickly.

The virus has the potential for rapid spread among humans. The rapid spread may result in an influenza pandemic. Influenza pandemics usually occur three to four times each century. The worst pandemic in the twentieth century was the 1918–19 Spanish flu. This pandemic resulted in an estimated 50 million deaths worldwide and is considered to be the deadliest plague in history. The flu appeared to have an avian origin. The critical change that made the influenza virus so infectious in people appears to have been a change in a single amino acid in the structure of the virus.

In 1918 an avian flu infected humans, and within months it adapted to its new host. The first of its new hosts were soldiers in World War I. The virus spread rapidly from person to person and killed more than any other plague in history (Fig. 5.3). In six months it killed over 30 million people.

There are two avenues for a new influenza pandemic to emerge. First, a dormant human flu virus may resurface. Because of the relatively long time period since the last outbreak of the virus, no natural defense mechanism would be available to protect against infection. Second, a nonhuman virus, such as the bird flu virus, may acquire the ability to infect humans and spread rapidly.

If the bird flu virus infected a human who is infected with a human flu virus, these two viruses might recombine into a new mutant, part human and part bird virus. Once established, the new flu virus would spread rapidly. If the bird flu virus spread to pigs, the virus would probably transfer more quickly and more easily to humans. This is because of the genetic similarities between humans and pigs.

The bird flu virus A(H5N1) is one of 15 known types of avian flu viruses. It typically is carried by ducks and does infect birds and pigs. Infected birds pass the virus in their feces and oral secretions for at least 10 days. The virus can leap from birds to humans, and viral infection in humans may result in death.

The A(H5N1) virus has infected millions of chickens in Cambodia, China, Indonesia, Japan, Pakistan, South Korea, Taiwan, Thailand, and Vietnam. Human cases of bird flu have been reported in China (Hong Kong), Thailand, and Vietnam.

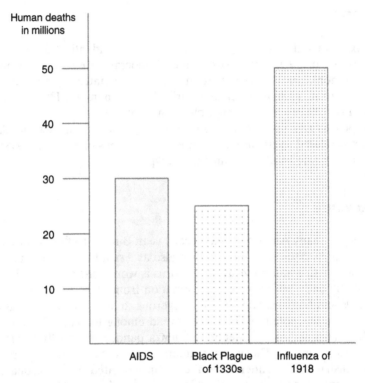

Figure 5.3 *Major plagues in world history. The influenza virus of 1918 is the worst plague to affect the human population. The virus was responsible for the death of over 50 million people worldwide. The plague caused by this virus killed more people than any other plague in history.*

The first reported infection of bird flu virus in humans was in Hong Kong in 1997.

Evidence suggests that humans become infected with bird flu virus through contact with infected birds. Therefore, individuals who come in contact with infected birds are at greatest risk of contracting bird flu. No evidence suggests human-to-human transmission of the bird flu virus.

Millions of chickens are being destroyed in many Asiatic countries to prevent the spread of the bird flu virus. Bird flu virus appears to be resistant to the older anti-viral medications amantadine and rimantadine. However, the virus appears to be sensitive to the newer antiviral medications, neuraminidase inhibitors Tamiflu and Relenza. The World Health Organization (WHO) is already working on a proto-type vaccine for bird flu virus.

6

Hepatitis Virus Group

Hepatitis is a disease characterized by a viral infection and inflammation of the liver. Scarring of the liver or cirrhosis may occur from this infection. Cases of viral hepatitis are reportable to the Centers for Disease Control and Prevention.

The viruses that cause hepatitis are collectively known as the hepatitis virus group. This group includes hepatitis A, B, C, D, E, and F and several lesser-known viruses (Table 6.1). Each hepatitis virus may produce complications (Table 6.2). The common types of viral hepatitis are A, B, and C, and viral hepatitis A is the most common. Hepatitis B and C viruses are bloodborne pathogens.

Although numerous routes of transmission are recorded for the hepatitis viruses, especially hepatitis A, hepatitis B, and hepatitis C (Table 6.3), approximately one-third of hepatitis infections result from unknown sources. The large number of infections from unknown sources indicates that infected individuals may or may not be members of "high-risk" groups (Table 6.4).

Because wastewater personnel have the opportunity for direct contact with fecal material that may contain hepatitis A (HAV), wastewater personnel could be at increased risk for HAV. Because of direct contact with wastewater, lack of personal protective equipment, or delay in prompt hand washing, wastewater personnel have an elevated concern for an increased risk of infection by HAV.

HAV is an important enteric pathogen. The importance or concern for this enteric pathogen is illustrated by several factors. These factors include (1) the severity of the disease that HAV causes, (2) the large number of viruses released and the long duration of release from infected individuals, and (3) the stability of the virus outside its host. HAV as well as all hepatitis viruses may be found in the wastewater as freely dispersed viruses, cell-associated viruses, or viruses associated with solids. When associated with cells or solids the viruses are more resistant to inactivation

Wastewater Pathogens, by Michael H. Gerardi and Mel C. Zimmerman
ISBN 0-471-20692-X   Copyright © 2005 John Wiley & Sons, Inc.

TABLE 6.1 Significant Members of the Hepatitis Virus Group

Viral Member	Acronym	Transmission	Vaccine Availability
A	HAV	Person-to-person (fecal-oral) Contaminated food or water	Yes
B	HBV	Sexual Via infected blood and body fluids Contact with contaminated surfaces From mother to child during birth or after	Yes
C	HCV	Injected drug use Via blood	No
D	HDV	Via blood Only in presence of hepatitis B	Yes (for HBV)
E	HEV	Fecal-oral in developing countries	No
F	HFV	Unknown	No

TABLE 6.2 Complications Associated with the Significant Members of the Hepatitis Virus Group

Viral Member	Complications
A	Fulminant hepatitis Relapse
B	Fulminant hepatitis Chronic liver disease Cirrhosis Liver cancer
C	Chronic liver disease Cirrhosis Liver cancer
E	High mortality in pregnant women

TABLE 6.3 Transmission Routes of the Most Common Hepatitis Viruses

Transmission Route*	Hepatitis Virus		
	A	B	C
Body piercing/tattooing		X	X
Contaminated food or water	X		
Contaminated surface		X	
Hemodialysis		X	X
Infected body fluids		X	
Intravenous (IV) drug use (shared needles)	S	X	X
Oral	X	U	S
Mother to child at birth		X	S
Needle stick injuries		X	X
Sexual, anal	X	X	S
Sexual, oral	X	X	S
Sharing earrings, toothbrushes, razors		X	
Transfusions	U	X	X

*X, definite; U, yes, but uncommon; S, suspected

TABLE 6.4 "High-Risk" Groups in Hepatitis A and Hepatitis B Infections

"High-Risk" Group	Hepatitis Virus	
	A	B
Day care centers	X	
Food service, food handlers	X	
Handlers of primates	X	
Health care workers		X
Hemodialysis patients		X
Homosexual men		X
Individuals with chronic liver disease	X	
Military personnel	X	
Multiple sex partners		X
Travelers to areas where sanitation is poor	X	
Wastewater operators and technicians	X	

TABLE 6.5 Significant Functions of the Liver

Function	Example
Carbohydrate metabolism	Regulates blood sugar level
Detoxification	Removes toxins from the blood
Filtration	Removes foreign bodies
Lipid metabolism	Synthesizes cholesterol
Protein metabolism	Deaminates amino acids; produces urea
Secretion	Secretes bile

by free chlorine [hypochlorous acid (HOCl) and hypochlorite ion (OCl$^-$)] and chloramine.

Hepatitis viruses are difficult to destroy. HAV is shed in high concentrations in the feces of infected individuals and can survive in moist environments for weeks to months. Hepatitis B (HBV) and hepatitis C (HCV) are highly resistant to destruction by detergents and elevated temperatures. HBV may survive for up to one month on dry surfaces and is 100 times more contagious than the human immunodeficiency virus (HIV). Also, HBV and HCV are not destroyed by alcohol.

LIVER

The liver is the largest organ in the human body. It performs numerous metabolic functions and serves as the body's filter for impurities. The liver is known as the "silent" organ. It suffers quietly through disease, that is, it is devastated slowly. Years of disease damage results in the loss of the liver's ability to perform its metabolic functions and remove impurities from the body (Table 6.5).

The liver is involved in carbohydrate, lipid, and protein metabolism. It helps to maintain an acceptable and safe level of blood sugar, oxidizes fatty acids, and produces lipoproteins, phospholipids, cholesterol, and fat molecules. The liver stores

TABLE 6.6   Flulike Symptoms

Fatigue
Fever
Headache
Lack of appetite
Muscle ache
Nausea
Joint stiffness
Vomiting

glycogen, minerals, and vitamins while destroying damaged blood cells and foreign bodies. It removes toxic substances including alcohol in the blood and secretes bile. The most important function of the liver is protein metabolism.

HEPATITIS

Hepatitis is an inflammation of the liver. There are several types of hepatitis, but the types have similar symptoms, including flulike symptoms (Table 6.6). The more distinctive symptoms of hepatitis include a rash, pain in the upper right quadrant of the abdomen, dark urine, and pale feces. The skin and sclera of the eyes turn yellow as hepatitis advances. The change in color is called jaundice and is caused by the accumulation of bile pigments.

Fulminant hepatitis occurs suddenly and severely and is associated with altered behavior. Liver and kidney failure and coma may accompany fulminant hepatitis. Hepatitis that lasts longer than six months is termed chronic.

Several types of viruses infect the liver and cause viral hepatitis (Fig. 6.1). Occasionally, the viral infection may lead to the development of nodules that can become cancerous. The viruses are distinguished by their route of infection and structural differences such as their protein coat.

Hepatitis A Virus

Hepatitis A virus (HAV) is highly contagious, and its major route of transmission is person-to-person by the fecal-oral route, for example, contact with foods or objects contaminated with the virus. Individuals infected with HAV may show no symptoms or may show flulike symptoms, dark urine, lightened stool, and jaundice. Individuals infected with HAV can be positively identified through blood tests.

HAV is a major waterborne disease in the United States. Outbreaks of hepatitis A due to contaminated drinking water have been reported.

Hepatitis B Virus

Hepatitis B virus (HBV) also is highly contagious, and its major route of transmission is by contact with HBV-contaminated bodily fluids such as blood, saliva, or semen. HBV enters the body through cuts and scrapes. Blood transfusions, hypodermic needles, or sexual activity may transmit HBV.

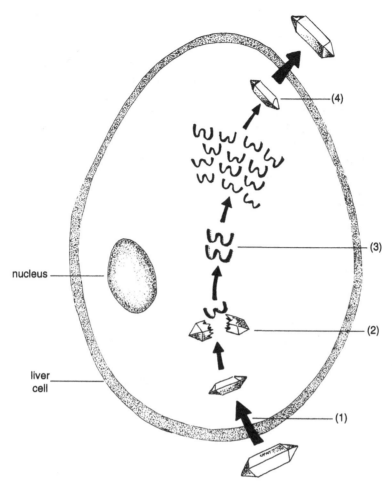

nucleus

liver
cell

(4)

(3)

(2)

(1)

Figure 6.1 *Viral infection and damage of liver cells. Liver cell infection and damage occurs when a hepatitis virus enters a liver cell (1) and releases its "genetic" material (2). The released "genetic" material takes over the reproduction functions of the liver cell nucleus and causes the liver cell to repli-cate copies (3) of the viral "genetic" material. Once replicated, the viral "genetic" material is packaged by the liver cell (4) and released from the cell. Once released, the liver cell is destroyed, and the new package of "genetic" material or virus is free to attack another liver cell.*

Individuals infected with HBV may show no symptoms or may show the same symptoms of hepatitis A. Although these symptoms may be more severe, most individuals infected with HBV recover. Chronic infections of HBV may result in cirrhosis, liver cancer, or liver failure. Individuals infected with HBV also can be identified through blood tests.

Hepatitis C Virus

Hepatitis C virus (HCV) is transmitted primarily in blood, for example, through shared razors or needles, from pregnant women to fetuses, or in blood transfusions or use of blood products. Individuals infected with HCV may show no symptoms or may show low-grade flulike symptoms. Individuals infected with HCV also can be identified through blood tests.

Like chronic hepatitis B, if left untreated chronic hepatitis C may result in cirrhosis, liver cancer, or liver failure. Liver failure due to chronic HCV infections is the leading cause of liver transplants in the United States.

HCV is the most problematic form of viral hepatitis. HCV is the deadly cousin to HAV and HBV. HCV may be responsible for about half of all cases of hepatitis. HCV constantly changes its outer coat, eluding the immune system and rendering vaccine development difficult. There is no vaccine for HCV, and a high percentage of hepatitis C cases result in chronic infections, cirrhosis, and hepatic cancer.

A comparison of major symptoms, transmission routes, treatment measures, vaccines, and risk groups for hepatitis viruses A, B, C, D, E, and F are presented in the following brief descriptions.

Hepatitis A (HAV)

- Risk groups

  Those living in areas with poor sanitation

  Those traveling in third world countries

  Those engaging in anal sex

  Intravenous drug users

- Symptoms

  Flulike illness with jaundice, nausea, fatigue, abdominal pain, loss of appetite, diarrhea, and fever

- Transmission routes

  Ingestion of feces-contaminated items

  Ingestion of contaminated ice, water, raw or partial cooked shellfish

  Ingestion of fruits and vegetables or uncooked food contaminated during handling

- Treatment

  Immune globulins 2 or 3 months before or 2 weeks after exposure

- Vaccine

  Vaccine is available

  Provides protection 4 weeks after first injection

  Several doses needed for long-term protection

Hepatitis B (HBV)

- Risk groups

  Intravenous drug users

  Those with multiple sex partners

  Those traveling or working in third world countries

  Those receiving blood transfusions before 1975 (HBV testing became available in 1975)

- Symptoms

Similar to hepatitis A
- Transmission routes
Contact with contaminated body fluids
Blood transfusions before 1975
Exposure to sharp objects containing contaminated blood
- Treatment
Hepatitis B immune globulin (HBIG) within 14 days after exposure
  Interferon α-2b and lamivudine
- Vaccine
Vaccines are available

Hepatitis C (HCV)

- Risk groups
Intravenous drug users
Those receiving blood transfusions before 1992
- Symptoms
Often no symptoms until liver damage occurs
Symptoms may be flulike
- Transmission routes
Blood-to-blood contact, especially intravenous drug use and needle sharing
Sexually transmitted
Blood transfusions before 1992
Exposure to items with contaminated blood
- Treatment
Interferon or combination drug treatments
Liver transplant for end-stage hepatitis C
- Vaccine
No vaccine available

Hepatitis D (HDV)

- Risk groups
Intravenous drug users
- Symptoms
Similar to hepatitis B
Occurs with HBV infection (HDV cannot survive on its own)
- Transmission routes
Occurs with the transmission of HBV
Sexual contact with HBV-infected individual
Exposure to sharp items with contaminated blood
- Treatment
Interferon

- Vaccine
  HBV vaccine provides protection

Hepatitis E (HEV)

- Risk groups
  Those traveling or working in third world countries
- Symptoms
  Similar to HAV
- Transmission routes
  Ingestion of feces-contaminated items
- Treatment
  No specific treatment available
- Vaccine
  No vaccine available

Hepatitis F (HFV)

Few cases of HFV have been described.

7

HIV

Infection with human immunodeficiency virus (HIV) results in progressive deterioration of the immune system. The deterioration is characterized by the selective destruction of T cells in the immune system and loss of generalized immune system activity. The final stages in the deterioration of the immune system—acquired immunity deficiency syndrome (AIDS)—results in opportunistic infections or malignancies. The first case of HIV in the United States reported by the Centers for Disease Control and Prevention was in June 1981.

AIDS occurs from the infection with the retrovirus HIV (Fig. 7.1). The average incubation period from infection with HIV to the development of AIDS is 8–10 years. However, AIDS can develop as quickly as 2 months and may take as long as 13 years to develop. AIDS is a preventable disease.

HIV is a bloodborne pathogen. Its modes of transmission are similar to those of the hepatitis B virus. Blood is the most important source of HIV and hepatitis B virus transmission in occupational settings. There is no evidence of an environmentally mediated mode of HIV transmission. Wastewater personnel are at minimal risk to HIV infection as long as they practice proper hygiene measures and use appropriate protective equipment. There are no known cases in which wastewater personnel became infected with HIV from contact with HIV-contaminated wastewater.

HIV does not have a fecal-oral route of transmission. HIV is most commonly transferred through heterosexual activity. There are several routes of transmission of HIV, including sexual activity and workplace accidents (Table 7.1). HIV has been found in many body fluids including blood and urine (Table 7.2). Sweat, bronchoalveolar lavage fluids, synovial fluid, feces, and tears have no or only low levels of HIV.

Because many HIV-infected body fluids such as blood, urine, and semen and wastes such as condoms that contain these fluids are discharged into wastewater

Wastewater Pathogens, by Michael H. Gerardi and Mel C. Zimmerman
ISBN 0-471-20692-X   Copyright © 2005 John Wiley & Sons, Inc.

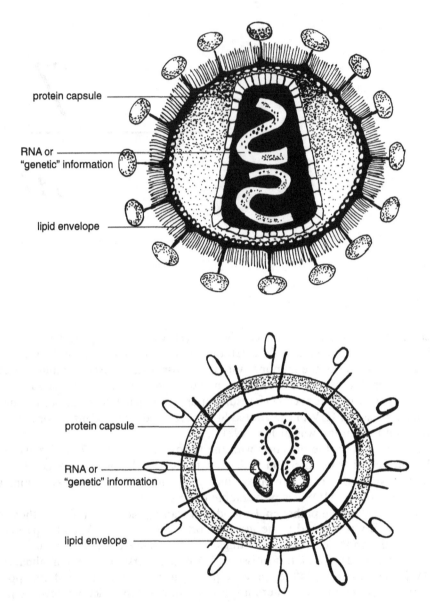

Figure 7.1 *Illustrations of HIV. The basic structural features of HIV are revealed in these two illustrations. The basic structure of HIV consists of a central core containing RNA or its "genetic" information that is surrounded by a protein capsule. The capsule contains numerous "extensions" or "door knobs" that are used for attaching to immune cells in humans. Surrounding the capsule is an additional protective feature, the lipid envelope.*

sewer systems, these substances contribute to the presence of HIV in wastewater. The greatest possible contribution to HIV levels in wastewater is semen. Discarded condoms potentially contribute the most to HIV levels. Contributions from mortuaries and menstrual blood may be important. Blood discharged by hospitals, blood banks, clinical laboratories, and medical and dental offices may contain HIV. HIV

TABLE 7.1   Routes of HIV Transmission

Anal sex
Exposure of health care workers to blood or bodily fluids from
    infected individuals resulting from contact with open wounds or
    accidental needle sticks
Heterosexual genital sex
Oral sex
Sharing of needles by intravenous drug users
Transfusion of contaminated blood
Uterotransmission to infants by infected mothers

**TABLE 7.2   Body fluids That Have Been Identified as
Containing HIV**

Blood
Blood products
Breast milk
Semen
Cerebrospinal fluid
Urine
Vaginal fluids

TABLE 7.3   Disinfectants That Inactivate HIV

Ethanol, 70%
Hydrogen peroxide, 0.3%
Isopropyl alcohol, 25%
Lysol, 0.5%
Quaternary ammonium chloride, 0.08%
Sodium hypochlorite, 0.1%

may be found in the wastewater as a suspended virus, a cell-associated virus, and adsorbed to solids.

Because of the routes of HIV transmission, no special precautions have been recommended to protect wastewater personnel from infection by HIV in wastewater. However, because of the presence of HIV in wastewater, wastewater personnel should follow appropriate hygiene measures and use appropriate protective equipment.

Although HIV is fairly stable in wastewater, it is a very fragile virus and quickly becomes inactive outside the human body. HIV is stable in wastewater for up to 12 hours, but its virulence or ability to cause successful infection is reduced steadily over these 12 hours.

HIV survival outside its host is affected by several factors, including temperature, pH, microbial activity, and whether the HIV is free, cell associated, or adsorbed to solids. HIV is very sensitive to acidic conditions, and free HIV gradually becomes inactive between pH 6.0 and 7.4. HIV is rapidly inactivated by desiccation and exposure to many commonly used disinfectants (Table 7.3).

HIV is easily inactivated by household bleach (supermarket hypochlorite-based bleach). Household bleach has been recommended to health care workers and intra-

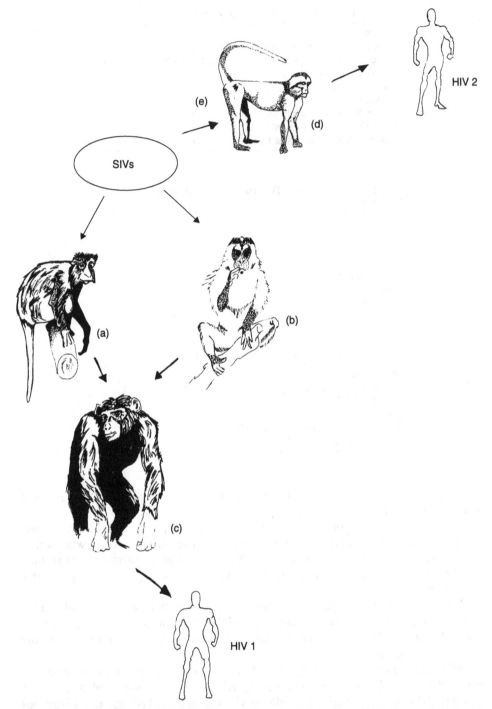

Figure 7.2 *Transfer of SIVs to humans. The presence of the AIDS virus or HIV in humans is believed to have occurred from the "jump" of the simian immunodeficiency viruses (SIVs) from primates to humans. Current information suggests that SIVs in the greater spot-nosed monkey (a) and the red-capped mangabey (b) "jumped" to the chimpanzee (c). From here SIV "jumped" to humans and mutated to form HIV 1. HIV 2 occurred from the mutation of the SIV virus in the sooty mangabey (d) when the virus "jumped" to humans.*

venous drug users as a way to inactivate HIV. Free HIV and cell-associated HIV are inactivated after 30-second exposure to undiluted bleach. If the bleach is diluted, its ability to inactive HIV is reduced. The sensitivity of HIV to disinfectants including chlorine makes it very unlikely that waterborne transmission of HIV will occur.

HIV may be rapidly inactivated in sewers by biological, chemical, and physical processes. Significant retention time promotes inactivation of HIV, and dilution in sewers reduces the concentration of HIV. Sewers simply are hostile environments for many pathogens including HIV.

There are different strains of HIV. These strains can spread and become resistant to drugs. Evidence suggests that two main families of HIV (HIV 1 and HIV 2) were introduced into humans several times.

Genetic evidence indicates that HIV originated from primates. The earliest sample of HIV-infected blood dates to 1959, and based on the mutation rate of HIV, it is estimated that HIV first appeared around 1930.

It is theorized that the AIDS virus in primates or simians (simian immunodeficiency virus or SIV) was passed from infected animals to humans (Fig. 7.2). SIV is widely found in primates, and when SIV jumped from primates to humans, HIV 1 and HIV 2 emerged. It is believed that SIV jumped from primates to humans during hunting and butchering of primates when human and infected primate blood mixed.

8

West Nile Virus

West Nile virus (WNV) is common in Africa, western Asia, and the Middle East and recently emerged in North America, where it has spread rapidly across the United States. The virus is carried by birds, particularly crows and jays, and has moved steadily south and west across the United States, following bird migration patterns.

The virus represents a significant threat to human and animal health and is considered to be a potentially lethal virus. The strain of WNV in the United States is very virulent, more virulent than those found in past decades. WNV is recognized as a cause of severe human meningoencephalitis inflammation of the spinal cord and brain.

Over 3500 human cases of WNV were reported in 44 of the contiguous United States during 2002. Only Arizona, Nevada, Oregon, and Utah did not report an avian, human, mammal, or mosquito infection of WNV.

WNV causes a variety of disease symptoms. The fatal form of the disease is encephalitis (inflammation of the brain). WNV is known also as Japanese flavivirus. WNV is one of six members of the Japanese encephalitis viral group within the family Flaviviridae (Table 8.1).

WNV is approximately 40–60 nm in size and is surrounded by an envelope. The virus is icosahedral and has a symmetric, near-spherical shape.

Members of the Japanese encephalitis viral group are arboviruses or arthropod-borne viruses that are carried by vectors such as mosquitoes. Arboviral encephalitises are zoonotic. These viruses are maintained in nature through their transmission between susceptible vertebrates (birds or small mammals) and blood-feeding arthropods (mosquitoes, sand flies, and ticks).

The life cycle of each virus in the Japanese encephalitis viral group is complex and incorporates a nonhuman, primary vertebrate host such as a bird or small

Wastewater Pathogens, by Michael H. Gerardi and Mel C. Zimmerman
ISBN 0-471-20692-X   Copyright © 2005 John Wiley & Sons, Inc.

TABLE 8.1 Members of the Japanese Encephalitis Viral Group

Viral Member	Geographic Distribution
Japanese encephalitis	Far East
Junjin	Pacific
Murray Valley encephalitis	Pacific
St. Louis encephalitis	North America and South America
Rocio	South America
West Nile virus	Africa, western Asia, Middle East, North America

mammal (Fig. 8.1). Therefore, the geographic distribution of each virus is associated with the location of its vectors, such as mosquitoes, and its hosts, such as birds. However, the virus can escape from its vector-host (mosquito-bird) life cycle and spread rapidly in an area in other vertebrates (cats, dogs, horses, or humans). This rapid movement of the virus often results in widespread mortality.

Humans, domestic animals, and other mammals can develop clinical illness or disease symptoms of WNV but usually are "incidental" or "dead-end" hosts, because they do not produce significant viremia (large numbers of viruses in their blood-stream). Because of insignificant viremia, humans and domestic animals do not contribute to the transmission cycle of WNV.

WNV was first isolated in 1937 in the West Nile district of Uganda and first documented in North America in 1999. The virus is transmitted to humans through mosquito bites. Mosquitoes capable of transmitting WNV include *Aedes*, *Anopheles*, *Culex*, *Mansonia*, and *Minomyia*. Mosquitoes most often associated with the transmission of the virus are members of the culicine group, primarily *Culex* species (Fig. 8.2 and 8.3). *Culex* is the genus most susceptible to infection with WNV. Mosquitoes are a food source for several animals including fish, birds, and bats.

Most recently, it has been observed that the virus may be transferred in humans through blood transfusions as well as breast milk and organ donations. Transfusion-related cases of WNV have been reported. Testing of the nation's blood supply began in July 2003 to prevent the transfer of WNV through blood transfusions.

WNV attacks the central nervous system. As a result of this attack, an individual may develop encephalitis or meningitis, if the brain or outer membrane around the central nervous system, respectively, is attacked.

An infected individual may or may not show disease symptoms of WNV. For example, an individual may be asymptomatic, that is, the individual is infected with the virus but displays no disease symptoms, or may be symptomatic, that is, the individual is infected with the virus and does display disease symptoms. The disease symptoms of infection from WNV may be acute or severe (Table 8.2).

Most individuals infected with WNV are asymptomatic. Approximately 20% of the individuals infected with WNV develop acute symptoms, and <1% of individuals infected with WNV develops severe symptoms. Most commonly occurring acute symptoms include body aches, mild fever, swollen glands, and rash. Acute symptoms may last a few days.

People who are elderly or have underlying health problems may get very sick and display severe symptoms. Most commonly occurring severe symptoms include disorientation, high fever, headache, muscle weakness, and neck stiffness. Severe symptoms may last several weeks, although neurological effects may be permanent.

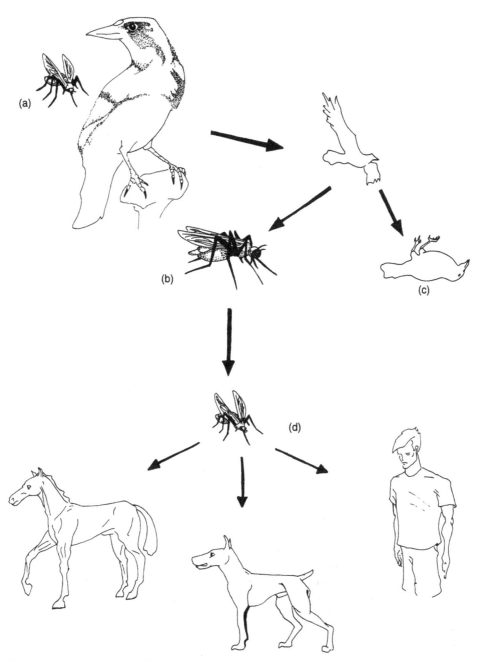

Figure 8.1 *Transmission of West Nile virus. There are several steps involved in the transmission of West Nile virus. An infected mosquito (a) bites a noninfected bird. The virus is transferred to the bird from the mosquito. The bird flies away, possibly to a new geographic area. The infected bird is bitten by a noninfected mosquito (b). The virus is transferred to the mosquito from the bird. The infected bird may die (c) or live and continue to serve as a vector for the spread and transmission of West Nile virus. The newly infected mosquito may bite another bird and also continue to spread West Nile virus, or the mosquito may bite a nonavian host such as a horse, dog, or human (d).*

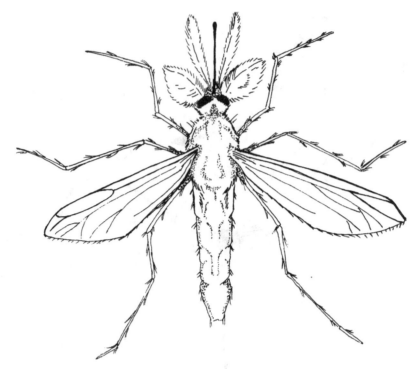

Figure 8.2 Culex sp.

The most severe symptoms of WNV are encephalitis, meningitis, and coma. Of the individuals who develop encephalitis, <0.01% die.

Individuals who develop disease symptoms from WNV infection usually display the symptoms within 3–15 days after infection. This period of time from infection to onset of disease symptoms is known as the incubation period.

All individuals within an area where the virus is active are at risk of becoming infected. However, individuals that are young, over 50 years of age, or have weakened immune systems have the highest risk of severe disease symptoms. The reason for the risk to individuals over 50 years of age is unclear at this time. Individuals who are elderly or have underlying health problems may get encephalitis. In some cases, individuals may slip into a coma and die.

There is no antiviral treatment for WNV. Treatment for individuals with severe disease symptoms includes hospitalization, use of intravenous fluids, respiratory support, and prevention of secondary infections. Infections confer lifelong immunity. However, this immunity weakens in later years.

Experimental work to control WNV infections consists of intravenous infusion of antibodies and the use of an antisense drug. Antibodies are substances produced by the immune system to fight infection. Antibodies produced by infected WNV individuals and given to individuals before exposure to WNV may act as a "passive" vaccine to prevent infection. The use of an antisense drug would make it impossible for the WNV to replicate in an infected individual.

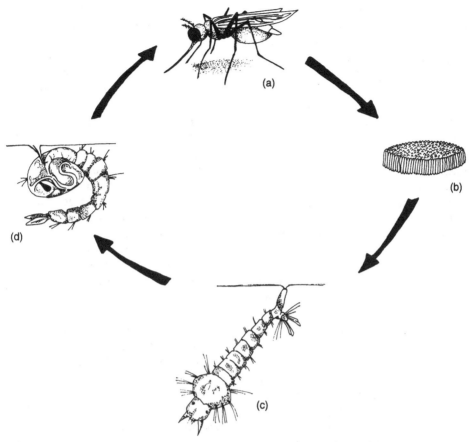

Figure 8.3 *Life cycle of* Culex pipiens. *There are four stages of development (life cycle) of the mosquito* Culex pipiens. *These stages include the adult (a), the eggs (b), the larva (c), and the pupa (d). The eggs, larva, and pupa are found floating on the surface of stagnant water.*

TABLE 8.2   Disease Symptoms of Infection from West Nile Virus

Acute Symptoms	Severe Symptoms
Body aches	Coma
Fever	Convulsions
Headache	Disorientation
Skin rash	Encephalitis
Swollen lymph glands	Headache
	High fever
	Limb weakness
	Meningitis
	Muscle weakness
	Neck stiffness
	Paralysis
	Stupor
	Tremors

There is no vaccine for humans at this time for WNV. However, several drug manufacturers are performing research work. Other research projects include new vaccines for horses, dogs, cats, and possibly birds.

Human infections with WNV are "reportable" to the Centers for Disease Control and Prevention (CDC). Over 3500 human cases of WNV were reported to the CDC during 2002. The number of cases in the United States in 2002 was the biggest reported outbreak of WNV in the world.

The virus is transmitted to humans through bites from infected mosquitoes. The mosquitoes carry the virus in their salivary glands. Mosquitoes become infected with the virus when they feed on infected birds. The population of viruses is amplified during periods of adult mosquito blood-feeding by continuous transmission between mosquitoes (vectors) and birds (reservoir hosts). Infected mosquitoes infect susceptible hosts during blood feeding. However, the virus may break free from the cycle of mosquitoes and birds and infect humans, horses, or small mammals. The virus cannot be passed from person to person or from bird or other animal to a person.

Mosquitoes feed on the blood of humans, birds, and other animals to obtain proteins that are needed for development of their eggs. Infected birds maintain an infectious viremia for 1–4 days after exposure. Most infected birds survive. After infection, birds develop lifelong immunity.

Adult mosquitoes live 7–30 days. Males usually live 7–10 days, whereas females live 30 days or more. Mosquitoes lay their eggs in stagnant water. The mosquitoes usually stay within 1 mile of the area where they are hatched and are active from May until late October. A wet spring leads to more WNV cases. At the end of mosquito season, the mosquitoes go into a "hibernation-like" state. The end of mosquito season occurs with the first hard freeze.

Mosquitoes may breed in any standing water that lasts more than 4 days. Although mosquitoes are active between dusk and dark, they are most active at dusk and dawn. Environmental conditions such as heavy rains followed by flooding and high temperatures can cause an increase in mosquito populations and an increase in the number of WNV cases.

West Nile virus is capable of infecting humans as well as numerous wild and domestic animals (alligators, bats, cats, crows, deer, dogs, eagles, great horned owls, hawks, horses, jays, mountain goats, pelicans, rabbits, and squirrels). West Nile virus is not fatal to all animals. Over time, some animals develop immunity.

West Nile virus has been detected in more than 100 different species of birds. Crows and jays are reported most commonly as being infected with WNV. There is no evidence that WNV can be transmitted by handling dead birds or any animal infected with WNV. However, proper precautions should be taken when handling any animal carcass. Gloves or double plastic bags should be used to place the carcass in the garbage, or the local health department should be contacted for guidance in disposing of the carcass.

Although cats and dogs can be infected with WNV, the virus usually does not cause extensive damage to these animals. Also, there is no evidence of animal-to-person transmission of WNV.

Horses can become infected with WNV. Approximately 40% of equine WNV cases result in death of the horse. During 2002 over 14,000 cases of equine WNV in 40 states were reported to the U.S. Department of Agriculture (USDA) and over 4300 horses were killed by the virus. A vaccine for horses has been developed.

TABLE 8.3   Personal Protective Measures

Apply insect repellant to thin clothing; mosquitoes can bite through thin clothing.
Apply insect repellant to expose skin; Centers for Disease Control and Prevention recommends repellants that contain DEET.
Stay indoors when mosquitoes are active (mainly dawn and dusk), especially in early evening hours.
Wear long-sleeved shirts and long pants.

TABLE 8.4   Control Measures

Aerate ornamental pools or stock them with fish that eat mosquitoes.
Clean roof gutters (clogged roof gutters can produce breeding grounds for mosquitoes).
Drain tank covers of water.
Drill holes in the bottom of containers such as recycling containers, left outdoors, so rainwater can drain out.
Install or repair screens.
Keep mosquitoes from entering buildings.
Repair leaky outdoor faucets.
Turn over wheelbarrows.
Use biological larvicides.
Use landscaping to eliminate standing water.

West Nile virus can be prevented in two ways. First, personal protective measures to reduce contact with mosquitoes should be practiced. Second, control measures should be used to reduce the population of infected mosquitoes.

Personal protective measures to reduce contact with mosquitoes (Table 8.3) include limiting outdoor activity between dusk and dawn, wearing protective clothing such as long sleeves and pants whenever practical, and wearing an insect repellant containing DEET (*N,N*-diethyl-*m*-toluamide). DEET should not be used on children under age 2, and only a 10% concentration should be used for children under age 10. Because mosquitoes can bite through thin clothing, an insect repellant should be applied to clothing. DEET does not kill mosquitoes. It repels them or makes an individual unattractive for feeding.

Control measures that can be used to reduce the population of infected mosquitoes in the environment (Table 8.4) include removing mosquito breeding sites by emptying or draining areas of standing water, such as buckets, ditches, gutters, sampling containers, and wheelbarrows, repairing screens, and using landscaping to eliminate low-lying areas. Chlorination of wastewaters also is helpful in controlling mosquito breeding sites. The addition of an insecticide or biological larvicide to wastewater to control mosquito breeding sites should be reviewed and approved by all appropriate regulatory agencies. In emergencies, wide-area aerial spraying is used to quickly reduce the number of mosquitoes. The U.S. Environmental Protection Agency may be contacted for selection of an appropriate insecticide.

Mosquitoes are attracted to people by skin odors and carbon dioxide from their breath. Insect repellants that contain DEET are the most effective in repelling pests like mosquitoes and ticks.

If you are outdoors when mosquitoes are active, it is important to apply an insect repellant that contains DEET. Repellants are only effective at short distances, and mosquitoes can find areas of the body that are not treated with DEET. Although DEET is found in different concentrations in repellants, a higher concentration of

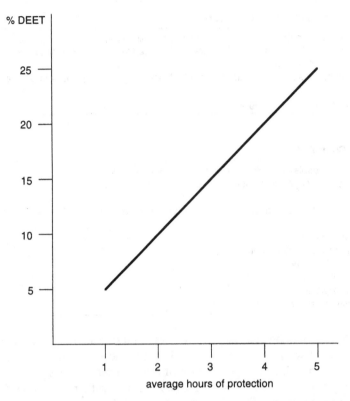

Figure 8.4   *% DEET in repellant and average hours of protection.*

DEET does not provide more protection from mosquito bites. A higher concentration provides a longer protection time (Fig. 8.4).

Insect repellants should be applied according to directions on the product label, and enough repellant should be applied to cover exposed skin and clothing. Repellants should not be applied to cuts or irritated skin, the eyes and mouth, and the hands of children. After returning indoors, the repellant should be washed from the treated areas.

State health departments employ a number of measures to control mosquitoes and outbreaks of WNV. These measures include mosquito surveillance and control, holding workshops to make people aware of WNV, and helping individuals to take appropriate control measures to prevent infestation from WNV. Reduction in mosquito populations is achieved by spraying mosquitoes and larvae and eliminating larval habitats.

Emerging concerns related to outbreaks of WNV are stormwater control and its role as a potential breeding ground for mosquitoes. Retention ponds often are used to collect runoff. The ponds are designed to drain slowly over several days. Because retention ponds drain slowly and mosquitoes breed anywhere they find standing water, ponds may need to be altered and designed to drain in less than 3 days. This time period is critical, because mosquitoes need 3 to 4 days to reproduce.

9

Bacteria

Bacteria are simple, unicellular organisms. Bacteria range in size from 0.1 to 15 μm, and the shape of most bacteria is either rod (bacillus), spherical (coccus), or spiral (spirillium) (Table 9.1).

Bacteria are ubiquitous in nature. They are found in water, soil, organic matter, and living bodies of plants and animals. They have a wide variety of nutritional requirements and may be autotrophic, parasitic, or saprophytic with respect to the manner in which they obtain their nutritional requirements (Table 9.2).

Most bacteria are harmless, and many colonize the human body, especially the digestive tract. They reproduce asexually, usually by splitting in half, and may be found as individual cells or as clusters or filaments (Fig. 9.1).

Bacteria are prokaryotic organisms or cells, whereas fungi, protozoa, and helminths are eukaryotic organisms (Fig. 9.2). Prokaryotic cells are generally smaller and less complex than eukaryotic cells and do not have a nucleus (a membrane-bound structure) that contains the genetic material of the cell. The genetic material of prokaryotic organisms is found in a large loop floating about in the cytoplasm of the cell. Basic components of bacterial cells include the cell wall, cell membrane, cytoplasm, ribonucleic acid (RNA), and ribosomes. Most bacteria are motile and move by means of flagella.

Bacteria are grouped according to several characteristics including cell shape, response to Gram staining (Fig. 9.3), and response to free molecular oxygen. Bacteria are surrounded by a rigid cell wall that provides protection. The cell wall is made of two layers of lipids surrounded by a sturdy carbohydrate capsule.

With few exceptions, the cell wall of bacteria reacts to the Gram stain. The Gram stain is a differential stain that is used to identify bacteria. Gram-positive bacteria have a thick cell wall made mostly of peptidoglycan. Gram-positive bacteria stain blue. Gram-negative bacteria have a thinner cell wall and stain red.

Wastewater Pathogens, by Michael H. Gerardi and Mel C. Zimmerman
ISBN 0-471-20692-X   Copyright © 2005 John Wiley & Sons, Inc.

TABLE 9.1   Shapes and Sizes of Bacterial Cells

Shape	Size, μm
Rod (bacillus)	0.5 to 1.0 width × 1.5 to 3 length
Spherical (coccus)	0.5 to 1.0 diameter
Spiral (spirillium)	0.5 to 5.0 width × 6.0 to 15.0 length

TABLE 9.2   Examples of Autotrophic, Parasitic, and Saprophytic Bacteria

Bacterial Type	Example
Autotrophic	*Nitrosomonas* spp. (nitrifying bacteria)
Parasitic	*Mycobacterium* (causative agent for tuberculosis)
Saprophytic	*Nocardia* (foam-producing filamentous organism)

Many antibiotics such as penicillin destroy specific bacteria by interfering with the construction of bacterial cell walls. Other antibiotics disrupt the functions of the cell membrane, inhibit protein synthesis, inhibit nucleic acid synthesis, or inhibit metabolic activity. Antibiotics that destroy bacteria may be nonspecific (broad spectrum) in nature and destroy a large diversity of bacteria or may be specific (narrow spectrum) in nature and destroy only a small diversity of bacteria (Table 9.3)

In addition to cell shape and cellular response to the Gram stain, cellular response to free molecular oxygen (O_2) is used to describe and identify bacterial cells (Table 9.4). There are three groups of bacteria according to response to free molecular oxygen.

Some bacteria, a minority, are aerobes. They can use only free molecular oxygen for cellular activity. Some bacteria, the most primitive are anaerobes. They cannot use free molecular oxygen for cellular activity. There are two types of anaerobes, oxygen tolerant and oxygen intolerant. Oxygen-tolerant anaerobes can survive in the presence of free molecular oxygen. Oxygen-intolerant anaerobes die in the presence of free molecular oxygen.

Most bacteria are facultative anaerobes and can carry out cellular activity with or without free molecular oxygen. Facultative anaerobes prefer free molecular oxygen, but they can use other molecules such as nitrate ions (NO_3^-), sulfate ions (SO_4^{2-}), and organic compounds to carry out cellular activity.

Bacteria are grouped or classified according to "Sections" rather than "Families" and "Orders." These Sections are listed and described in *Bergey's Manual of Determinative Bacteriology*. The Sections are determined in large part by bacterial cell shape, bacterial response to the Gram stain, and bacterial response to free molecular oxygen (Table 9.5).

Most bacteria are free living. However, there are bacteria that infect humans and animals. These bacteria are pathogenic organisms. Pathogenic bacteria usually enter a host through ingestion, inhalation, and invasion (Table 9.6).

Numerous, significant pathogenic bacteria are found in wastewater (Table 9.7). The most common bacterial pathogens found in raw wastewater are *Salmonella* and *Shigella*. *Escherichia coli* generally is not considered to be a true pathogen because

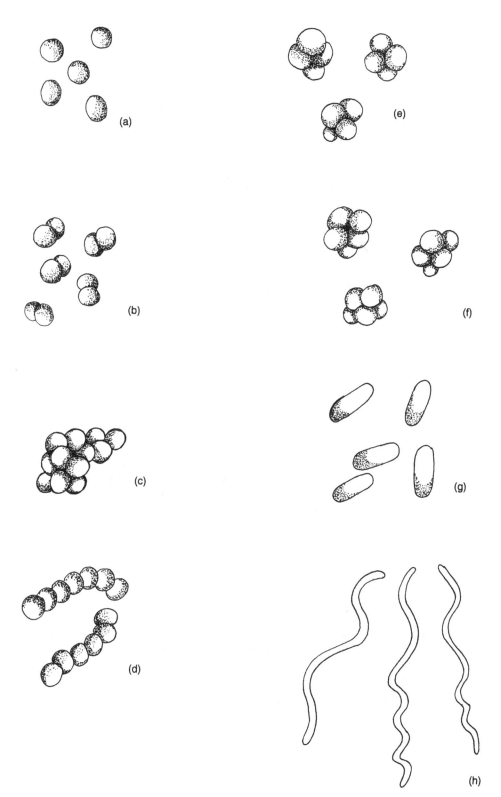

Figure 9.1 Patterns of bacterial growth. There are several common patterns of bacterial growth. These patterns include individual (a), pairs (b), irregular clusters (c), chains or filaments (d), groups of four or tetrads (e), and cubes or sarcinae (f). There are three basic cell shapes. These shapes are spherical or coccus (a), rod or bacillus (g), and spiral or spirillum (h).

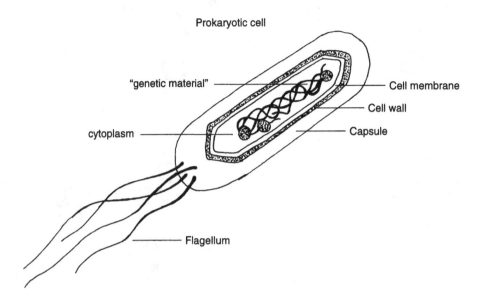

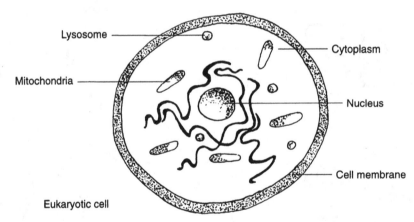

Figure 9.2 *Important features of prokaryotic and eukaroytic cells. Prokaryotic cells are simple in structure and cellular content compared with eukaryotic cells. Important features of the prokaryotic cells include a cell wall, cell membrane, cytoplasm, and "genetic" material of the cell that is loosely bundled as threads in the cytoplasm. There are no membrane-bound organelles in prokaryotic cells. Some prokaryotic cells may contain a flagellum or several flagella and a protective capsule. Important features of the eukaryotic cells include the cell membrane, mitochondria, lysosomes, and nucleus. The mitochondria, lysosomes, and nucleus are all membrane-bound structures. The nucleus contains the "genetic" material of the cell.*

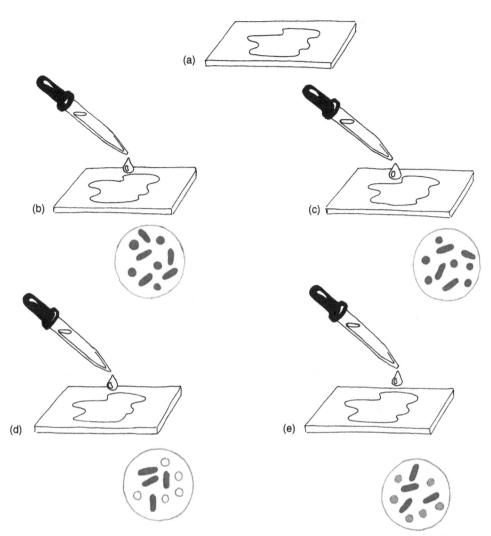

Figure 9.3 *Gram staining technique. Gram staining consists of the application of four reagents or solutions to a smear of bacteria. The bacteria in the smear respond to the solutions as Gram positive (blue) or Gram negative (red). In the Gram staining technique a smear of bacteria is made on a clean microscope slide (a). Once the smear has dried (bacteria fixed to the slide), crystal violet is applied to the smear (b). All bacteria appear blue after the crystal violet application. After crystal violet, a mordant or iodine is applied to the smear (c). All bacteria appear blue after the iodine application. Next, a decolorizing agent or alcohol wash is applied to the smear (d). The crystal violet-iodine complex is "washed" from the Gram-negative bacteria. After the alcohol wash, Gram-positive bacteria remain blue, whereas Gram-negative bacteria become colorless. To better observe the Gram-negative bacteria, a counterstain or safranin solution is applied (e). After the safranin application Gram-positive bacteria remain blue, whereas Gram-negative bacteria stain red.*

TABLE 9.3   Examples of Broad-Spectrum and Narrow-Spectrum Antibiotics

Bacteria Destroyed	Antibiotic	
	Broad Spectrum	Narrow Spectrum
Gram positive	Ampicillin	Erythromycin
	Gentamicin	Penicillin
Gram negative	Kanamycin	Polymyxins

TABLE 9.4   Groups of Bacteria According to Response to Free Molecular Oxygen

Group	Example	Disease
Aerobe	*Nocardia caviae*	Nocardiosis
Facultative anaerobe	*Escherichia coli*	Gastroenteritis
Anaerobe	*Clostridium tetani*	Tetanus

TABLE 9.5   Examples of Sections of Bacteria in *Bergey's Manual of Determinative Bacteriology*

Section	Bacterium in Section
Gram-negative aerobic rods and cocci	*Pseudomonas aeruginosa*
Gram-negative facultatively anaerobic rods	*Escherichia coli*
Gram-positive cocci	*Staphylococcus aureus*
Gram-positive, endospore-forming rods and cocci	*Clostridium tetani*
Gram-positive, irregular non-spore-forming rods	*Actinomyces israelii*

TABLE 9.6   Examples of Pathogenic Bacteria and Their Routes of Transmission

Bacterium	Disease	Route of Transmission
Bacillus cereus	Food poisoning	Ingestion
Bordetella pertussis	Pertussis or whooping cough	Inhalation
Campylobacter jejuni	Gasteroenteritis	Ingestion
Clostridium perfringens	Gangrene (gas gangrene)	Invasion
Clostridium tetani	Tetanus	Invasion
Legionella pneumophila	Legionellosis	Inhalation
Mycobacterium tuberculosis	Tuberculosis	Inhalation
Salmonella typhus	Typhoid fever	Ingestion
Staphylococcus epidermidis	Skin infection	Invasion
Streptococcus pneumonia	Pneumonia	Inhalation
Streptococcus spp.	Skin infection	Invasion
Vibrio cholerae	Cholera	Ingestion

TABLE 9.7 Pathogenic Bacteria Associated with Wastewater

Bacterium/bacteria	Disease
Actinomyces israelii	Actinomycosis
Campylobacter jejuni	Gastroenteritis
Clostridium perfringens	Gangrene (gas gangrene)
Clostridium tetani	Tetanus
Escherichia coli—enteroinvasive	Gastroenteritis
Escherichia coli—enteropathogenic	Gastroenteritis
Escherichia coli—enterotoxigenic	Gastroenteritis
Escherichia coli—enterohemorrhagic O157:H7	Gastroenteritis and hemolytic uremic syndrome
Leptospira interrogans	Leptospirosis
Mycobacterium tuberculosis	Tuberculosis
Nocardia spp.	Nocardosis
Salmonella paratyphi	Paratyphoid fever
Salmonella spp.	Salmonellosis
Salmonella typhi	Typhoid fever
Shigella spp.	Shigellosis
Vibrio cholerae	Cholera (Asiatic cholera)
Vibrio parahaemolyticus	Gastroenteritis
Yersinia enterocolitica	Yersiniosis (bloody diarrhea)

it is a normal inhabitant of the gastrointestinal tract. However, because *Escherichia coli* is an opportunistic pathogen and is found in very high numbers in the gastrointestinal tract and wastewater, it is included with the significant pathogenic bacteria. *Leptospira interrogans* and *Campylobacter jejuni* perhaps are the most significant. Genera of bacteria that contain pathogenic species and are of some concern to wastewater personnel include *Achromobacter*, *Aeromonas*, *Bacillus*, *Enterococcus*, *Erwinia*, *Legionella*, *Pseudomonas*, *Shigella*, *Staphylococcus*, and *Streptococcus*.

There are two types of pathogenic bacteria. "True" pathogens such as *Shigella* spp. and *Vibrio cholerae* are aggressive and are transmitted from person-to-person and contact with animals and their wastes. "Opportunistic" pathogens such as *Aeromonas hydrophilia*, *Escherichia coli*, *Mycobacterium avium*, and *Pseudomonas aeruginosa* are typically found on or in the human body and do not cause disease unless the body's immune system is weakened by injury, a "true" pathogen, or physiological disease.

The infective dose for most bacteria, especially the enteric bacteria, for a healthy individual typically is greater than 10,000 viable cells. However, the infective doses for some bacteria, such as *Salmonella* and *Shigella*, are much lower.

In addition to "true" and "opportunistic" bacterial pathogens, some bacteria secrete toxins that are capable of causing disease. Toxins secreted by bacteria consist of endotoxins and exotoxins.

Several pathogenic bacteria produce unique cellular structures that protect them from the effects of chemicals, desiccation, heat, and other environmental factors such as pH change. These structures include endospores (Fig. 9.4) and capsules (Fig. 9.5).

Bacteria in the genera *Bacillus* and *Clostridium* encase their genetic material and critical cellular components into tiny internal packages or endospores (Table 9.8). Endospores or spores are highly resistant to harsh environmental conditions and

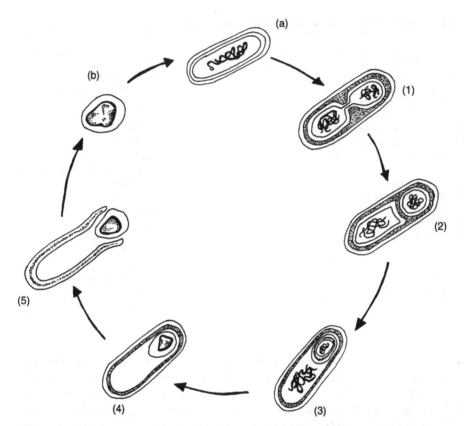

Figure 9.4 *Endospore formation. There are several steps that must be completed for an endospore-forming bacterium (a) to form an endospore (b). First, the genetic material or DNA or the cell must duplicate and the duplicated DNA must be separated from other cellular components by a septum (1). Next, the duplicated DNA must be isolated and surrounded by its own cytoplasm and membrane (2). Spore formation begins to materialize when the septum surrounds the duplicated DNA (3). Once the septum surrounds the duplicated DNA, a coat begins to develop around the duplicated DNA (4). The developing endospore at this time is referred to as a forespore. Once coat development is complete, the spore is released (5).*

permit the bacteria to survive adverse conditions such as high temperatures and desiccation. The presence of dipicolinic acid and a large quantity of calcium ions (Ca^{2+}) within the spores contributes to their resistance. Spores are formed during harsh environmental conditions. When conditions are favorable, the spores germinate and release active or vegetative cells.

Some bacteria such as *Streptococcus pneumoniae* are able to survive attack by a host's defenses by forming a capsule. The capsule is a protective structure that is located outside the cell wall of the bacterium. Capsules typically consist of complex polysaccharides packaged as a loose gel.

The capsule interferes with the host's ability to perform phagocytosis (engulfing and destroying foreign bodies such as pathogenic bacteria). The capsule contains toxic compounds that destroy phagocytes or white blood cells in the host's body. In addition to *Streptococcus pneumoniae*, examples of capsule-forming bacteria include *Yersinia pestis* (*Pasteurella pestis*), the infectious agent of plague, and *Bacillus anthracis*, the infectious agent of anthrax. *Bacillus anthracis* is both a spore-

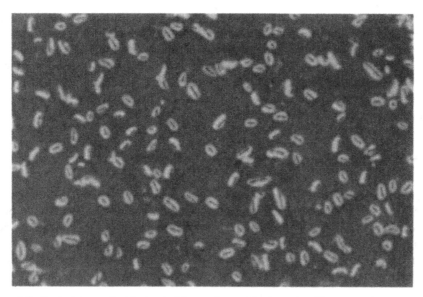

Figure 9.5 *Bacterial capsule. The bacterial capsule provides bacteria with some resistance to harsh environmental or operational conditions. Under the capsule stain, the bacteria cells stain whereas the capsule does not stain. The capsule appears as a "halo" around the bacteria.*

TABLE 9.8   Examples of Spore-Forming Bacteria

Bacterium	Disease
Bacillus anthracis	Anthrax
Bacillus cereus	Food poisoning
Clostridium botulinum	Botulism
Clostridium perfringens	Gas gangrene
Clostridium tetani	Tetanus

TABLE 9.9   Examples of Enzymes Produced by Pathogenic Bacteria

Toxin	Function
Coagulases	Clot blood
Collagenase	Breaks down collagen in connective tissues of muscles
Hemolysins	Destroy red blood cells
Hyaluronidase	Destroys polysaccharides that bind cells together
Leukocidins	Destroy white blood cells
Streptokinase	Dissolves blood clots

forming and a capsule-forming bacterium. The spore is formed outside the host, and the capsule is formed inside the host.

In addition to the production of spores and capsules, many pathogenic bacteria produce enzymes and toxins to overcome a host's defenses. A variety of enzymes are produced that attack specific defenses (Table 9.9). Enzymes and toxins permit the spread of infection by pathogenic bacteria.

Several pathogenic bacteria are becoming resistant to antibiotic treatment. Hard-to-treat skin infections once common to hospitals and correctional institutions are now occurring in athletes. *Staphylococcus aureus* is an antibiotic-resistant bacterium. The bacterium is resistant to treatment by penicillin-related antibiotics, including methicillin. Symptoms of methicillin-resistant *Staphylococcus aureus* (MRSA) infections include fever, pus, swelling, and pain. MRSA can progress to life-threatening blood and bone infections.

BACTERIAL PATHOGENS AND THEIR DISEASES

Presented here are brief descriptions of the wastewater bacterial pathogens of concern to wastewater personnel and their diseases. The descriptions of the bacteria include basic characteristics, disease, occurrence, reservoir, and typical mode of transmission. Perhaps the two most important bacterial pathogens are *Campylobacter jejuni* and *Leptospira interrogans*. *C. jejuni* is reviewed in more detail after the bacterial descriptions, whereas *L. interrogans* is reviewed in Chapter 10.

Actinomyces israelii

Characteristics: Gram-positive filament

Disease: actinomycosis

Disease identification: a chronic disease most often localized in the jaw, thorax, or abdomen

Occurrence: sporadically throughout the world

Reservoir: humans (oral cavity)

Mode of transmission: person-to-person

Clostridium perfringens

Characteristics: Gram-positive rod

Disease: food poisoning

Disease identification: intestinal disorder characterized by sudden onset of abdominal colic followed by diarrhea

Occurrence: worldwide

Reservoir: soil, gastrointestinal tract of humans and animals

Mode of transmission: ingestion of contaminated food

Clostridium tetani

Characteristics: Gram-positive rod

Disease: tetanus

Disease identification: acute disease induced by the bacteria that grow anaerobically at the site of a deep puncture wound or injury and produce neurotoxin (tetanospasmin). Disease is characterized by painful muscular contractions, primarily of the masseter and neck muscles.

Occurrence: worldwide

Reservoir: intestinal tract of animals, soil, feces-contaminated environments

Mode of transmission: tetanus spores introduced into the body during injury

Escherichia coli—Enteroinvasive

Characteristics: Gram-negative rod

Disease: gastroenteritis

Disease characteristics: usually localized in the colon. Symptoms consist of fever and diarrhea, occasionally bloody diarrhea.

Occurrence: contaminated food and water outbreaks in communities, areas of poor sanitation

Reservoir: infected individuals

Mode of transmission: fecal contamination of food and water

Escherichia coli—Enteropathogenic (ETEC)

Characteristics: Gram-negative rod

Disease: gastroenteritis

Disease characteristics: "classical" acute diarrhe disease in newborn nurseries

Occurrence: contaminated food and water outbreaks in communities, areas of poor sanitation

Reservoir: infected individuals

Mode of transmission: fecal contamination of food and water

Escherichia coli—Enterotoxigenic

Characteristics: Gram-negative rod

Disease gastroenteritis

Disease characteristics: similar to *Vibrio cholerae*—profuse watery diarrhea without blood or mucus; abdominal cramps, vomiting, acidosis, and dehydration

Occurrence: contaminated food and water outbreaks in communities, areas of poor sanitation

Reservoir: infected individuals

Mode of transmission: fecal contamination of food and water

Escherichia coli—enterohemorrhagic O157:H7

Characteristics: Gram-negative rod

Disease: Gastroenteritis, hemolytic uremic syndrome (kidney damage)

Disease characteristics: diarrhea, abdominal cramps, fever, and vomiting

Occurrence: contaminated food and water outbreaks in communities, areas of poor sanitation

Reservoir: infected individuals

Mode of transmission: fecal contamination of food and water

Mycobacterium tuberculosis

Characteristics: Gram-positive rod

Disease: tuberculosis

Disease characteristics: first appears as lesions on the lungs; may progress to pulmonary tuberculosis or meningeal or other extrapulmonary involvement including intestinal tract, bone nervous system and the skin.

Occurrence: worldwide

Reservoir: primarily humans, in some areas diseased cattle

Mode of transmission: exposure to bacilli in airborne droplet from sputum of infected individuals. Bovine tuberculosis results from exposure to diseased cattle or ingestion of unpasteurized milk or dairy products.

Nocardia spp., including N. asteroides, N. caviae, and N. brasiliensis

Characteristics: Gram-positive filament

Disease: Nocardiosis

Disease characteristics: often originating in the lungs and spreading to produce abscesses of brain, subcutaneous tissue, and other organs

Occurrence: sporadically throughout the world

Reservoir: soil

Mode of transmission: inhalation of contaminated particles

Salmonella spp.

Characteristics: Gram-negative rod

Disease: Salmonellosis (gastroenteritis)

Disease characteristics: gastroenteritis; sudden onset of abdominal pain, diarrhea, nausea, fever, and sometimes vomiting. Dehydration may be severe. Anorexia and looseness of the bowels often persist for several days.

Occurrence: worldwide; more extensively reported in North American and European countries; often classified with food poisoning

Reservoir: humans, domestic and wild animals

Mode of transmission: ingestion of contaminated food

Salmonella paratyphi

Characteristics: Gram-negative rod

Disease: paratyphoid fever

Disease characteristics: enteric infection with abrupt onset, continued fever, enlargement of the spleen, sometimes rose spots on trunk, usually diarrhea, and involvement of lymphoid tissues of the mesentery and intestines

Occurrence: sporadically or in limited outbreaks

Reservoir: humans

Mode of transmission: direct or indirect contact with feces or urine of patient or carrier

Salmonella typhi

Characteristics: Gram-negative rod

Disease: typhoid fever

Disease characteristics: sustained fever, headache, malaise, anorexia, enlargement of the spleen, rose spots on the trunk, cough, constipation, and involvement of the lymphoid tissue

Occurrence: worldwide

Reservoir: humans

Mode of transmission: direct or indirect contact with feces or urine of patient or carrier

Shigella spp.

Characteristics: Gram-negative rod

Disease: Shigellosis (bacillary dysentery)

Disease characteristics: edema, superficial ulceration of the large intestine, bleeding, watery or bloody diarrhea, fever, drowsiness, anorexia, nausea, and abdominal pain

Occurrence: worldwide

Reservoir: humans

Mode of transmission: person-to-person by fecal-oral route by inanimate objects, contaminated food or water

Vibrio cholerae

Characteristics: Gram-negative, slightly curved rod

Disease: cholera

Disease characteristics: enterotoxins released cause sudden severe nausea, vomiting, and abdominal pain, copious diarrhea; most deaths related to shock

Occurrence: Asia, eastern Europe, and North Africa

Reservoir: human, possible environmental reservoirs

Mode of transmission: contaminated water

Vibrio parahaemolyticus

Characteristics: Gram-negative rod

Disease: gastroenteritis, vibriosis

Disease characteristics: watery diarrhea, abdominal cramps, nausea, vomiting, fever, and headache

Occurrence: worldwide, especially Japan

Reservoir: marine and coastal environments

Mode of transmission: ingestion of raw or inadequately cooked and contaminated seafood

Yersina enterocolitica

Characteristics: Gram-negative rod

Disease: gastroenteritis, yersiniosis

Disease characteristics: release of enterotoxin causes severe abdominal pain (similar to appendicitis), low-grade fever, headache, pharyngitis, anorexia, and vomiting.

Occurrence: worldwide

Reservoir: animals, especially avian and mammalian

Mode of transmission: fecal-oral transmission with infected persons or animals

CAMPYLOBACTER JEJUNI

Curved, Gram-negative rod bacteria in the genus *Campylobacter* cause campylobacteriosis. *Campylobacter jejuni* is the principal species associated with campylobacteriosis. Campylobacteriosis is the most common gastroenteritis in the United States and is associated with abdominal pain, fever, and bloody diarrhea. In addition to gastroenteritis, species of the genus *Campylobacter* also cause dental disease and systemic infections of the brain, heart, and joints. Campylobacteriosis occurs throughout the United States, and its occurrence peaks during the summertime. Campylobacteriosis can be treated with antibiotics.

Species of the genus *Campylobacter* are not common inhabitants of the intestinal tract of humans. They are carried by numerous animals, including cats, cattle, dogs, and sheep. Transmission of *Campylobacter* spp. usually is by ingestion of contaminated food, milk, and water. Unlike many other pathogenic bacteria, *Campylobacter jejuni* does not reproduce in food.

10

Leptospira Interrogans

Leptospirosis is a disease of concern to wastewater personnel, especially sewer system personnel. Leptospirosis is caused by several species of bacteria in the genus *Leptospira*. The most frequently identified bacterium associated with the disease is *Leptospira interrogans*.

The bacteria responsible for leptospirosis are Gram-negative spirochetes (Fig. 10.1). The spirochetes live in the kidneys of mammals, especially rodents, and are shed in urine. Viable spirochetes are found in not only the wastewater but also the biofilm lining the inside of sanitary manholes. Infections of sewer system personnel with *Leptospira* spp. have occurred.

Symptoms of leptospirosis include fever, headache, jaundice, and kidney and liver damage. Death can occur with severe infections of *Leptospira* spp. The incubation period of leptospirosis is 10–12 days. Most cases of leptospirosis are asymptomatic, and recovery from most cases usually takes 2–3 weeks. The spirochetes are susceptible to treatment by a variety of antibiotics, if the antibiotics are taken in the early stages of infection.

Many untreated cases of leptospirosis result in death. The most virulent form of leptospirosis is known as Weil disease and is associated with jaundice and liver damage.

Transmission of the spirochetes can occur through several means. Contact with urine from infected animals is a major means of transmission of leptospirosis. In many parts of the world, over 50% of the rodents are infected with *Leptospira* spp. and serve as carries of the disease. Transmission also can occur through the consumption of contaminated food or water. Many wild animals, including rodents, as well as cats and dogs carry the spirochetes. Vaccines for cats and dogs are available.

Urine (human and animal) that is contaminated with spirochetes may be discharged directly to the sewer system or indirectly to the sewer system through inflow

Wastewater Pathogens, by Michael H. Gerardi and Mel C. Zimmerman
ISBN 0-471-20692-X Copyright © 2005 John Wiley & Sons, Inc.

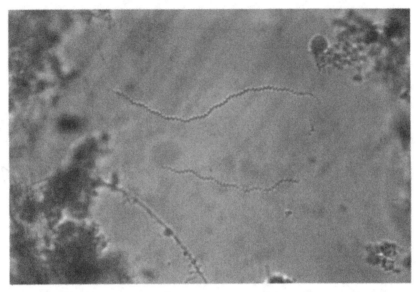

Figure 10.1   *Spirochetes. Spirochetes are single-celled bacteria that have a rigid and spiral or wave-like shape and are very motile. Spirochetes vary greatly in length from 5 to 15 μm in size. There are free-living and pathogenic spirochetes.*

and infiltration (I/I) after rainfall. Contaminated urine also is directly discharged to the sewer system from rodents that inhabit the sewer system.

The spirochetes gain access to the host by crossing the mucous membranes of the eyes, nose, and mouth, through skin abrasions, or through the consumption of contaminated food or water. Once inside the host, the spirochetes enter the convoluted tubules of the kidneys. Here, they multiply, causing disease, and are released in urine.

Outside the host, the spirochetes die quickly in acidic water. They can survive for approximately 3 months in neutral or slightly alkaline water. If successful, the spirochetes enter a new host.

Reduction in the risk of exposure to *Leptospira* spp. can be achieved through rodent control in sanitary sewers and destruction of spirochetes in the biofilm within sanitary manholes. *Leptospira* spp. is destroyed in wastewater in the presence of detergents. Suspect contaminated biofilm should be sprayed with muracic acid or hydrochloric acid (HCl). The decrease in pH in the biofilm produced by the application of HCl results in destruction of the spirochetes. After treatment, the biofilm should be hosed down with water.

11

Fungi

Fungi are a diverse group of organisms. Important characteristics that are used to classify or group fungi are their means of reproduction and their life cycles (Table 11.1). Fungi such as molds and mushrooms are multicellular, whereas some, such as yeast, are unicellular.

Fungi are saprophytes. They obtain their nourishment from dead organic matter or living organisms. However, fungi are not obligate parasites, because all fungi can obtain their nourishment from dead organisms. When fungi infect a living organism, they kill cells and obtain their nourishment as saprophytes from dead cells.

There are fewer than 50 fungal species that cause disease in humans. Most pathogenic fungi associated with wastewater are opportunistic. Fungal infections or mycoses are either superficial or systemic. Superficial mycoses occur on the hair, nails, and skin, whereas systemic mycoses commonly occur in the respiratory tract.

Candidiasis is a superficial mycosis caused by *Candida albicans*. Aspergillosis is a systemic mycosis caused by *Aspergillus fumigatus*. These two fungi, especially *Aspergillus fumigatus*, are of particulate concern to wastewater personnel.

Wastewater personnel at composting operations are exposed to *Aspergillus fumigatus* as well as *Blastomyces* spp. and *Histoplasma* spp. These fungi cause respiratory tract disease and runny nose. *Aspergillus fumigatus* is the causative agent for aspergillosis or "farmer's lung." The disease is a chronic, debilitating allergic lung disease.

ASPERGILLUS FUMIGATUS

Aspergillus fumigatus (Fig. 11.1) is an important mold fungal pathogen. *A. fumigatus* as well as other species of *Aspergillus* are important pathogens of special concern

Wastewater Pathogens, by Michael H. Gerardi and Mel C. Zimmerman
ISBN 0-471-20692-X   Copyright © 2005 John Wiley & Sons, Inc.

TABLE 11.1 Major Groups or Phyla of Fungi

Common Name	Phylum	Example	Significance
Bread molds	Zygomycota	*Rhizopus*	Some are opportunistic human pathogens, e.g., *Rhizopus*.
Club fungi	Basidiomycota	*Claviceps*	Includes mushrooms, rusts, smuts, and toadstools. Some mushrooms, e.g., *Amanita*, produce toxins. Some are opportunistic human pathogens, e.g., *Cryptococcus*.
Sac fungi	Ascomycota	*Neurospora*	Many are of industrial and medical value, e.g., *Penicillium* used for antibiotic production and *Saccharomyces* used for wine production and leavening bread, *P. roquefortii* and *P. camemberli* used for color, flavor, and texture production of Roquefort and Camembert cheeses
Water molds	Oomycota	*Saprolegnia*	Includes mildews and plant blights; Many are plant parasites including fungus responsible for the Irish potato famine in the 1840s

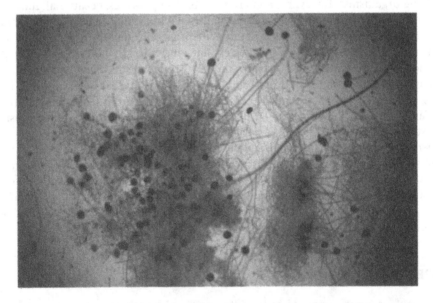

Figure 11.1 Aspergillus fumigatus. *The fungus* Aspergillus fumigatus *is associated with thermophilic composting operations and is responsible for severe respiratory tract infections. The fungus has a rootlike or mycelia growth that contains ball-like structures or conidia (spores). The inhalation of the conidia may result in an infection of* Aspergillus fumigatus.

to wastewater composting operators. The diseases include allergic reactions, colonization in immunocompetent individuals, and systemic infection. Concerns related to diseases caused by *A. fumigatus* are heightened by the lack of effective therapy.

Aspergillus fumigatus is a saprophytic fungus that is commonly found in the soil. Here, *A. fumigatus* performs an important beneficial role in recycling carbon and nitrogen by contributing to the decomposition of organic matter.

Aspergillus fumigatus reproduces through the production of spores or conidia. The conidia are gray-green in color and small in size (2.5–$3.0\,\mu m$ in diameter). The small size of the conidium allows it to reach the alveoli of the lungs. Here, infection of the lungs may occur, resulting in aspergillosis.

Part III

Protozoans and Helminths

12

Introduction to Parasitic Protozoans and Helminths

Along with beneficial organisms that degrade and remove wastes, wastewater contains many viruses, pathogenic bacteria and fungi, and parasites. Parasites consist of unicellular protozoans and multicellular worms or helminths (roundworms and flatworms). Parasites feed off other living organisms and cause disease. Individuals infected with parasites routinely shed these organisms into sewer systems through fecal waste. Many of the parasites thrive and even multiply in wastewater. Infections with parasites usually occur when an individual swallows parasites or their protozoan cysts or oocysts and helminth eggs. Some parasitic infections such as that caused by the hookworm (*Necator* sp.) (Fig. 12.1) occur when the parasite burrows through the skin.

Wastewater personnel have potentially greater risk for parasitic infection than the general population. Many individuals infected with parasites as well as viruses and pathogenic bacteria and fungi show no symptoms of infection or disease. These individuals are considered "subclinical" or asymptomatic.

Complex life cycles and specialized structures allow some parasites to be more resistant to wastewater treatment systems than other. The number and variety of parasites in wastewater is a reflection of the diseases in the community. The destruction or survival of parasites in wastewater treatment systems is due to numerous factors. However, no inclusive statement can be made with respect to how many or what species will survive a treatment system.

A parasite is an organism that lives on (ectoparasite) or in (endoparasite) another species. This species is the host. The parasite normally does not kill its host, because the life of the parasite also would be terminated. Hosts are placed in different groups.

The host in which the parasite reaches sexual maturity and reproduction is termed the definitive host. If no sexual reproduction occurs in the life cycle of the

Wastewater Pathogens, by Michael H. Gerardi and Mel C. Zimmerman
ISBN 0-471-20692-X   Copyright © 2005 John Wiley & Sons, Inc.

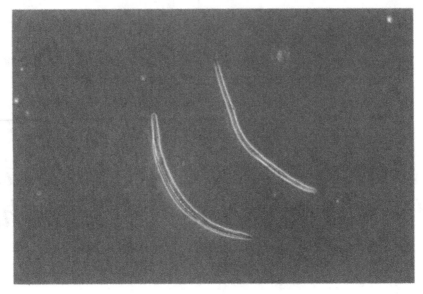

Figure 12.1   Hookworms, Necator *sp.*

parasite, such as that for the protozoan *Giardia lamblia*, the host that is considered to be the most important is the definitive host.

In an intermediate host some development of the parasite occurs. An example of parasite development would be the maturation of larval stages in an intermediate host. However, the parasite does not reach sexual maturity in the intermediate host. An intermediate host or paratenic host is useful or necessary for completion of the life cycle of the parasite.

Parasites may be obligate parasites or facultative parasites. Most parasites are obligate parasites, that is, they spend at least part of their lives as parasites in order to survive and complete their life cycle. However, many obligate parasites have free-living stages outside a host, including some period of time as a protective egg or cyst. The eggs and cysts of various parasites often are found in wastewater and sludge.

Facultative parasites normally are not parasitic but can become so when they are accidentally consumed or enter a wound or other body opening. An accidental or incidental parasite is a parasite that enters or attaches to a host that is not its definitive or intermediate host.

Broad taxonomic groups (phyla) that may have eggs or cysts in wastewater or sludge include Protozoa (Table 12.1), Nematoda (Table 12.2), Platyhelminthes (Table 12.3), and Acanthocephla (Table 12.4). Protozoa examples include *Entamoeba* and *Giardia* (Fig. 12.2). Nematoda (roundworms) examples include *Ascaris* and *Trichuris*. Platyhelminthes (flatworms) examples include flukes such as *Fasciola* and tapeworms such as *Hemenolepis*. An example of Acanthocephla (spiny headed worms) is *Macacanthorhynchus*. The worm phyla (Nematoda, Platyhelminthes, and Acanthocephala) generally are considered to be the "helminths."

TABLE 12.1 Parasitic Protozoans Found in Wastewater and Sludge

Protozoans	Probable Identity	Definitive Host
Balantidium	*Balantidium coli*	Mammals, birds
Capillaria spp.	*Capillaria hepatica*	Rats
	Capillaria gastrica	Rats
	Capillaria spp.	Wild mammals, birds
Coccidia oocysts	*Isospora* spp.	Dogs, cats
	Eimeria spp.	Mammals, birds
Cruzia-like eggs	*Cruzia americana*	Opossums
Cryptosporidium	*Cryptosporidium parvum*	Mammals, birds, reptiles, fish
	Cryptosporidium spp.	Mammals, birds, reptiles, fish
Entamoeba	*Entamoeba histolytica*	Humans
Giardia	*Giardia lamblia*	Humans
	Giardia spp.	Mammals
Naegleria	*Naegleria fowleri*	Humans
	Naegleria gruberia	Mammals
Toxoplasma	*Toxoplasma gondii*	Cats
Trichosomoides-like	*Trichosomoides crassicauda*	Rats
eggs	*Anatrichosoma buccalis*	opossums

TABLE 12.2 Parasitic Nematodes Found in Wastewater and Sludge

Nematode	Probable Identity	Definitive Host
Ascaris eggs	*Ascaris lumbricoides**	Humans
	Ascaris suum*	Pigs
Ascaridia-like eggs	*Ascaridia galli*	Domestic poultry
	Heterakis gallinae	Domestic poultry
Enterobius	*Enterobius vermicularis*	Humans
Gongylonema-like eggs	*Gongylonema neoplasticum*	Rat
	Gongylonema pulchrum	Cattle, pigs
Hookworms	*Anclyostoma* sp.	Humans
	Necator americanus	Dogs
Parascaris	*Paracaris equorum*	Horses
Strongyloides	*Strongyloides stercoralis*	Humans, dogs
Toxocara	*Toxocara canis*	Dogs
	Toxocara cati	Humans, cats
Toxascaris-like eggs	*Toxascaris leonina*	Dogs
Trichinella	*Trichinella spiralis*	Humans, pigs
Trichuris whipworms	*Trichuris trichiura*	Humans
	Trichuris suis	Pigs
	Trichuris vulpis	Dogs

* Eggs of *A. lumbricoides* and *A. suum* are indistinguishable.

TABLE 12.3   Parasitic Platyhelminths Found in Wastewater and Sludge

Platyhelminth	Probable Identity	Definitive Host
Diphyllobothrium-like eggs	*Diphyllobothrium latum*	Humans, dogs, bears
	Diphyllobothrium spp.	Dogs, bears, birds
Echinococcus	*Echinocococcus granulosus*	Rats, mice, dogs
	Echinococcus multilocularis	Humans
Fasciola	*Faciola* sp.	Humans
Hymenolepis diminuta eggs	*Hymenolepis diminuta*	Rats
Hymenolepis nana eggs	*Hymenolepis nana**	Humans, rats
Hymenolepis sp. eggs	*Hymenolepis* spp.	Birds
Schistosoma	*Schistosoma* sp.	Humans
Schistosoma mansoni	*Schistosoma mansoni*	Humans
Spirometra-like eggs	*Spirometra mansonoides*	Cats, dogs
Taenia sp. eggs	*Taenia solium*	Humans, pigs
	Taenia saginata	Humans
	Taenia pisiformis	Cats
	Hydratigera taeniaeformis	Dogs

* A genus name change of *Vampirolepis nana* for *Hymenolepis nana* has been suggested.

TABLE 12.4   Parasitic Acanthocephalans Found in Wastewater and Sludge

Spiny-Headed Worm	Probable Identity	Definitive Host
Acanthocephalan eggs	*Macracanthorhynchus hirudinaceus*	Pigs

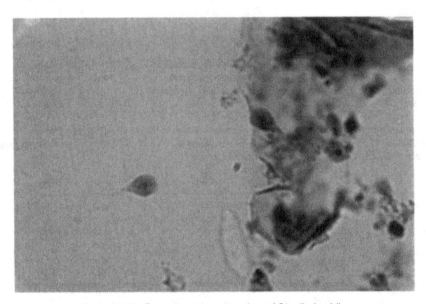

Figure 12.2   *Pear-shaped trophozoites of* Giardia lamblia.

13

Protozoans

The term "protozoan" is a common name of single-celled, eukaryotic organisms that are either animal-like, fungus-like, or plant-like (Table 13.1). Protozoans also can be distinguished or grouped by their inability or ability to move with cilia (ciliates), flagella (flagellates), or pseudopodia (amoebae). Protozoans that have no direct locomotive ability are coccidians.

Ciliates possess numerous, hair-like structures or cilia that are found in rows on the surface of the organism. The cilia beat to provide locomotion and direct a water current of substrate (bacteria) to the mouth opening. Ciliates are very beneficial in wastewater treatment systems and perform numerous roles in the stabilization of wastes. An important role performed by ciliates is the consumption of numerous, dispersed bacteria from the bulk solution in the activated sludge process. Only one ciliate, *Balantidium coli*, is known to parasitize the human intestinal tract.

Flagellates possess whip-like structures or flagella. The beating action of the flagella provides the protozoan with locomotion. There are numerous species of flagellates that parasitize the human intestinal tract. The most common flagellate in the human intestine is *Giardia lamblia*.

Amoebae move by a streaming motion of their cytoplasm (jellylike, cellular content) against a flexible cell membrane. The movement of the cytoplasm and the cell membrane appears as a slowly extending "false foot" or pseudopodia.

Amoebae commonly are found in polluted water and in the intestines of humans and many animals. Although most amoebae are harmless, certain amoebae such as *Entamoeba histolytica* can cause dysentery. This protozoan disease can be fatal.

Coccidia are all parasitic on animals and humans. The coccidian *Toxoplasma gondii* causes toxoplasmosis that is particularly dangerous to pregnant women. Another coccidian of medical importance is *Isospora belli*. This protozoan is responsible for coccidiosis.

Wastewater Pathogens, by Michael H. Gerardi and Mel C. Zimmerman
ISBN 0-471-20692-X   Copyright © 2005 John Wiley & Sons, Inc.

TABLE 13.1   How Protozoans Satisfy Substrate (Food) Requirements

Protozoan Group	Means for Satisfying Substrate Requirements
Animal-like	Ingestive heterotrophs that engulf food and digest food within the cell
Fungus-like	Absorptive heterotrophs that excrete enzymes outside the cell and digest food
Plant-like	Photosynthesis

TABLE 13.2   Major Waterborne Parasitic Protozoans

Organism (locomotion)	Disease (site affected)	Major Reservoir
Balantidium coli (cilia)	Dysentery/intestinal ulcers (gastrointestinal tract)	Human feces
Cryptosporidium (coccidium)	Cryptosporidiosis (gastrointestinal tract)	Human and animal feces
Entamoeba histolytica (pseudopodia)	Amoebic dysentery (gastrointestinal tract)	Human feces
Giardia lamblia (flagella)	Giardiasis (gastrointestinal tract)	Human and animal feces
Naegleria graberi* (pseudopodia)	Amoebic meningoencephalitis (central nervous system)	Soil and water

* *Naegleria* is not a parasite, but a free-living protozoan pathogen.

Although most protozoans are free living in soil or water, some protozoans can be parasitic. Parasitic protozoans are small in size (2–200 μm). The animal-like protozoans contain several parasites of concern to wastewater personnel including *Cryptosporidium*, which only recently has been recognized as a major cause of diarrhea (Table 13.2).

The form of a protozoan parasite that lives inside the host is called the trophozoite stage. Most protozoan parasites produce a cyst stage that can survive outside their hosts under adverse environmental conditions. Encystment can be triggered by factors such as lack of nutrients or accumulation of toxic metabolites but is usually caused by a response of the parasite to the host's immune system. Encystment also is a natural part of the parasite's life cycle.

CRYPTOSPORIDIUM PARVUM

Cryptosporidium parvum is a coccidian protozoan that is known to infect humans and most animal species (e.g., calves, cats, chickens, dogs, lambs, mice, and turkeys). The life cycle of *Cryptosporidium parvum* (Fig. 13.1) contains of an infective thick-walled oocyst (5–6 μm in diameter). An infected individual may release up to 10,000,000,000 oocyts per day. After ingestion by a suitable host, the oocysts undergo excystation and release infective sporozoites that parasitize epithelial cells in the gastrointestinal tract. The parasite causes a profuse and watery diarrhea that typically lasts 10–14 days in normally healthy individuals. There is no drug therapy to control this parasite, and the infection may be fatal in immunodeficient individuals.

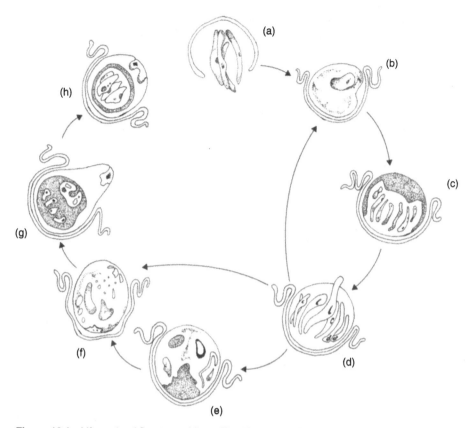

Figure 13.1 *Life cycle of* Cryptosporidium. *The life cycle of* Crytosporidium *involves several stages of development. When a host swallows an infective oocyst, sporozoites emerge from the oocyst in the intestinal tract of the host (a). Each sporozoite is capable of invading an epithelial cell of the intestinal tract by attaching to the surface of a cell. The cell then grows around the sporozoite. Here, the sporozoite feeds on the host cell and causes disease. As the sporozoite feeds, it matures. The mature sporozoite is referred to as a schizont (b). The schizont divides to form eight merozoites (c). The banana-shaped merozoites escape from the epithelial cell (d) and may invade another epithelial cell that continues the infection or may develop into a male reproductive cell—the microgametocyte (e) or a female reproductive "egg"—the macrogametocyte (f). The male reproductive cell produces "sperm cells" or microgametes—which burst out of the epithelial cell and invade an epithelial cell that contains a female reproductive cell. When the "sperm cells" fuse with the egg, fertilization occurs, and an immature oocyst is formed (g). When the oocyst matures (h), it develops a thick outer wall and is released from the epithelial cell into the lumen of the small intestine, where it is incorporated in fecal waste.*

ENTAMOEBA HISTOLYTICA

Entamoeba histolytica forms infective cysts (10–15 μm in diameter) that are shed for relatively long periods by asymptomatic carriers. The cysts persist well in wastewater and may subsequently be ingested by new hosts. Levels of cysts in wastewater may be as high as 5000 per liter.

The protozoans is transmitted to humans primarily by contaminated water and food. It infects the large intestines and causes amebiasis or amoebic dysentery.

Symptoms vary from diarrhea alternating with constipation to acute dysentery. It may cause ulceration of the intestinal mucosa, resulting in secondary infection.

GIARDIA LAMBLIA

Giardia lamblia is the causative agent of giardiasis and is a cosmopolitan protozoan parasite. Its life cycle (Fig. 13.2) consists of a trophozoite, resembling a "badminton

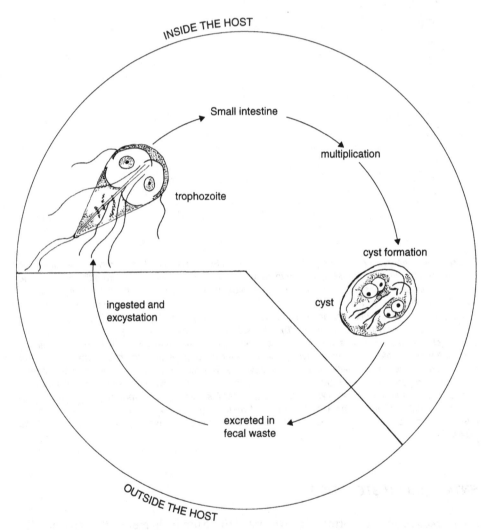

Figure 13.2   *Life cycle of* Giardia lamblia. *Infection by* Giardia lamblia *begins when viable cysts are ingested. Inside the digestive tract of the host, the cysts undergo excystation. Excystation results in the release of active trophozoites that attach to the wall of the small intestine, where multiplication of the trophozoites occurs. As trophozoites leave the host in fecal waste, many undergo cyst formation. These cysts can survive for several weeks outside the host, whereas trophozoites cannot survive outside the host.*

racket" measuring 10–20 μm long and the infective cyst stage, which measures 15 μm long by 5 μm wide, complete with a thick wall and four nuclei. Domestic wastewater is a significant source of *Giardia lamblia*, and wild and domestic animals act as important reservoirs of cysts. Levels of cysts in wastewater may be as high as 1,000,000 per liter.

Giardia lamblia has an incubation period of 1–8 weeks and causes diarrhea, abdominal pain, nausea, fatigue, and weight loss. *Giardia lamblia* is very resistant to chlorine concentrations typically used to treat final effluents at wastewater treatment plants.

VERIFICATION OF INFECTION

If an infection of a parasitic protozoan is suspected, its verification involves examination of stool (fecal smear for intestinal parasites) or sputum (for lung parasites) samples to detect trophozoites or cysts. Wastewater personnel use common sense, proper hygiene measures, and appropriate protective equipment to avoid protozoan infection.

MONITORING FOR PROTOZOANS

Because of their small size, monitoring for parasitic protozoans in wastewater can be tedious and often requires filtering large volumes of samples followed by hours of microscopic analysis. A further complication is that if a cyst is detected, there is no easy way to determine whether it is still viable or infective. New protocols and standard techniques involving immunofluoresence microscopy are currently being developed. However, these techniques may prove too costly for routine monitoring.

14

Helminths

The term "helminth" comes from the Greek for "worm." Although there are many species of free-living worms, there are thousands of species of parasitic helminths. There are two groups of helminths. These groups are the flatworms and round-worms. Flatworms consist of tapeworms (cestodes) and flukes (trematodes). Round-worms also are known as nematodes and include the spiny-headed, acanthocephalan worms.

Helminths exist in two forms. The first form is an actively growing larva or worm. The larva is found inside the definitive host and produces eggs or ova. The egg or ovum is the second form and leaves the host in fecal waste. The ovum is or develops a protective structure that is resistant to adverse conditions and has the ability to infect a new host.

In the life cycle of a helminth, there may be one or more intermediate hosts. In the intermediate host the larva or ovum (Fig. 14.1) must survive the acid condition of the host's digestive tract in order to multiply before becoming infective. In some life cycles, humans can be the intermediate host to the larva. An example of a larval infection in humans is the cysticerci larva of the pork tapeworm, *Taenia solium* (Fig. 14.2).

The helminths include several worms that are important parasites to humans, animals, and crops. Many helminths are transmitted through human and animal wastes. Major parasitic helminths of concern to wastewater personnel are listed in Table 14.1.

Most surveys of wastewater parasites have simply determined the presence or absence of parasitic stages. However, these surveys have not determined whether the stages are infective. Wastewater and sludge in the United States contain the following principle species of parasites (in order of occurrence): *Ascaris* sp.,

Wastewater Pathogens, by Michael H. Gerardi and Mel C. Zimmerman
ISBN 0-471-20692-X   Copyright © 2005 John Wiley & Sons, Inc.

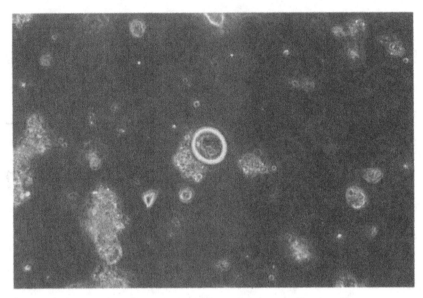

Figure 14.1 *Ovum or egg of the pork tapeworm,* Taenia solium.

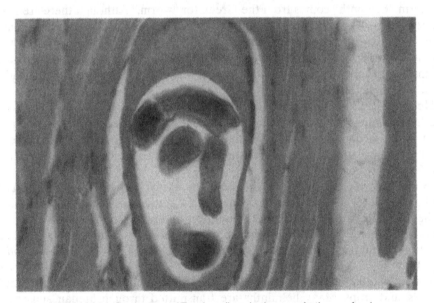

Figure 14.2 Cysticercus *of* Taenia solium *in the muscle tissue of swine.*

TABLE 14.1   Major Parasitic Helminths of Concern to Wastewater Personnel

Helminth Group	Organism	Disease
Cestodes (tapeworm)	*Taenia saginata*	Beef tapeworm
	Taenia solium	Pork tapeworm
Hookworms	*Ancyclostoma duodenale*	Hookworm
	Necator americanus	Hookworm
Nematodes (roundworms)	*Ascaris lumbricoides*	Ascariasis
	Trichuris trichiura	Trichuriasis (whipworm)
Trematodes (flukes)	*Schistosoma mansoni*	Schistosomiasis

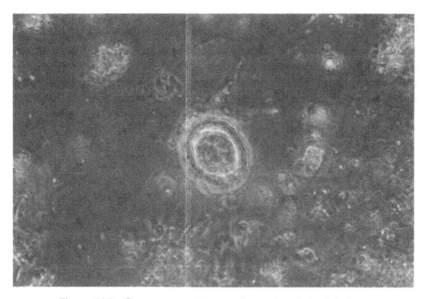

Figure 14.3 *Ovum or egg of the roundworm* Ascaris lumbricoides.

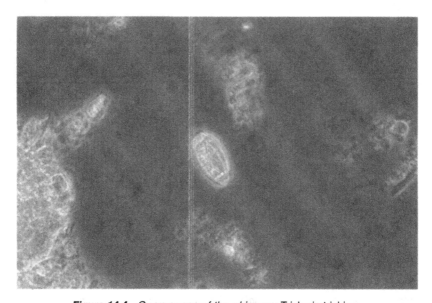

Figure 14.4 *Ovum or egg of the whipworm* Trichuris trichiura.

Toxocara sp., *Trichuris* sp., *Hymenolepis* sp., *Capillaria* sp., and *Taenia* sp. In suitable environments, the eggs of *Ascaris* (Fig. 14.3), *Trichuris* (Fig. 14.4), and *Toxocara* can survive in soil for several years. The eggs of *Ascaris* can remain infective up to seven years.

ASCARIS LUMBRICOIDES

Ascaris lumbricoides is a large, cosmopolitan, parasitic nematode. The female measures 20–25 cm in length, and the male measures 12–31 cm in length. The host swallows embryonated eggs with second-stage larva to initiate infection (Fig. 14.5). On reaching the small intestine, the second-stage larva is stimulated to hatch. It then

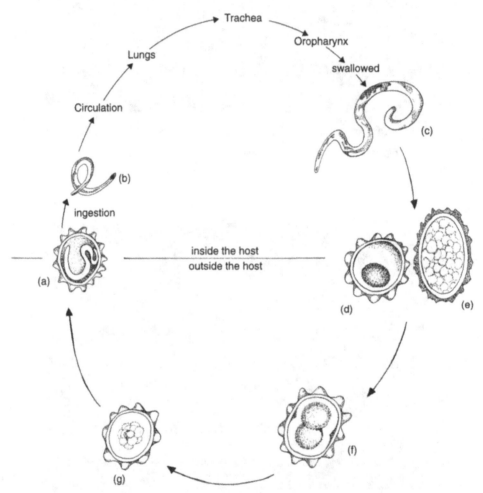

Figure 14.5   *Life cycle of Ascaris lumbricoides. The life cycle of Ascaris lumbricoides begins when developed (embryonated) ova or eggs (a) are ingested by the host. In the small intestine of the host the young Ascaris worms or larvae hatch from the ova (b). The larvae penetrate the intestinal wall and enter the circulatory system. The circulatory system carries the larvae to the heart and lungs. In the lungs the larvae leave the capillaries of the circulatory system. During the migration of the larvae through the circulatory system, the larvae undergo growth and development through a series of molts. From the lungs the larvae move into the bronchi and are swallowed into the digestive tract. Here, the larvae molt their last time and become sexually mature worms (c). The mature worms are capable of developing nonfertile eggs (d) and fertile eggs (e) that leave the host in its fecal waste. The fertilized eggs undergo development (f–g) outside the host, until infective larvae are developed in the ova (a).*

rapidly penetrates the intestinal wall of the host and enters the circulatory system. The larva develops to a third-stage worm during a 1- to 2-week migratory period. The third-stage larva eventually is pumped to the pulmonary artery, penetrates the respiratory tract (trachea) to the pharynx, and is swallowed by the host. After two more molts in the small intestine, the larva becomes fully mature and begins to reproduce.

The average life span for an adult *Ascaris* is 2.5–3 years, and once mature the female can produce over 200,000 fertilized unembryonated eggs per day. The eggs normally embryonate in soil and take 2–3 weeks to do so. The fertilized egg is ovoid and measures 60–70 mm × 30–50 mm in size. The unfertilized egg is larger.

The thick shells of *Ascaris* ova (Fig. 14.3) prevent the penetration of toxic materials and provide resistance to many adverse environmental conditions including changes in temperature. For this reason, the survival of ascarid ova often has been considered to be an indicator for measuring the efficiency of wastewater treatment processes in destroying pathogens. Sludge usually is not an ideal environment for ascarid ova to become infective embryonated eggs. However, additional research must be performed to determine ascarid egg embryonation in sludge.

HYMENOLEPIS (VAMPIROLEPIS) NANA

Hymenolepis nana is a small, cosmopolitan parasitic tapeworm (Fig. 14.6). The adult worms range in size from 30 to 40 mm in length. Infections of *Hymenolepis nana* occur after ingestion of embryonated eggs (Fig. 14.7) that contain the hexacanth or onchosphere embryo (Fig. 14.8). The onchosphere embryo penetrates the tissues of the small intestine and develops into a cysticercoid larva. Maturation to the adult tapeworm takes approximately 30 days from the time of ingestion of the egg

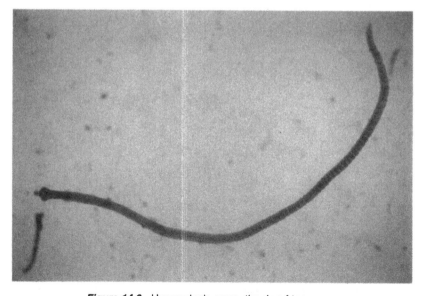

Figure 14.6 Hymenolepis *nana, the dwarf tapeworm.*

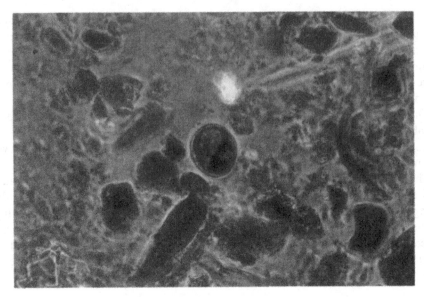

Figure 14.7 Hymenolepis nana, *embroynated egg.*

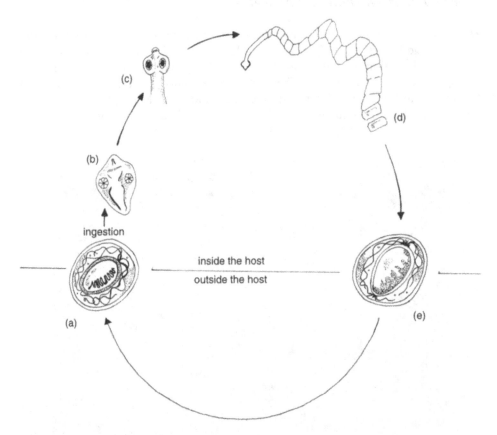

Figure 14.8 *Life cycle of* Hymenolepis nana. *The life cycle of* Hymenolepis nana *begins when a rat or other host (a) ingests embryonated eggs. In the intestinal tract of the host, onchospheres (b) hatch from the eggs and develop into cysticercoids in the villa of the small intestine. The cysticercoids emerge from the villa as mature worms (c) and attach to the wall of the small intestine. Embryonated eggs (e) are released by the proglottids (d) of the worm and leave the host in fecal waste.*

(Fig. 14.9). The life cycle of *Hymenolepis nana* also can be completed in small rodents such as mice and rats. Therefore, these rodents or reservoir hosts can provide for an increase in the number of *Hymenolepis nana* eggs in wastewater.

TAENIA SOLIUM

Taenia solium is a large, parasitic tapeworm that is found in all parts of the world where pork or pork products are consumed. Adult worms live in the intestinal tract of its host and may exceed 3 m in length. The oval eggs of *Taenia solium* are 30–40 μm in diameter and contain a thick, resistant coat. Often, the oncosphere is visible inside the egg.

In the normal life cycle of *Taenia solium*, humans develop the tapeworm after eating poorly cooked pork that is infected with the cysticercus larva of the worm (Fig. 14.10). Therefore, wastewater and sludge are not the usual vectors for infection of the adult worm. However, direct ingestion of *Taenia solium* eggs by humans can cause an infection known as cysticercosis (*Cysticercus cellulosae*). Depending on the location of the larva, this condition can be serious. As in *Hymenolepis nana*, the adult tapeworm cannot develop from this larva infection in humans.

TRICHURIS TRICHIURA

Trichuris trichiura is known commonly as the whipworm because the anterior portion of the worm is elongated and thin and the posterior portion of the worm is fleshy and bulbous. This structural feature of the worm resembles a "whip." *Trichuris trichiura* is a small, cosmopolitan, parasitic roundworm of the large intestine of its host. The worm is especially prevalent in warm climates. The adult female is 35–50 mm in length, and the adult male is 30–45 mm in length.

Infection occurs through the ingestion of embryonated eggs (Fig. 14.11). The eggs contain the first-stage larvae that hatch in the small intestine. The larvae are carried into the large intestine and penetrate the epithelium of the intestine where they molt four times before becoming adult worms. The adults begin to produce eggs about 1–2 months after infection. Female worms shed about 3000–5000 fertilized unembryonated eggs per day and may continue to do so for up to 2 years. Like *Ascaris lumbricoides*, the eggs of *Trichuris trichiura* contain a resistant shell and require 2–3 weeks in soil to become infective. The eggs of *Trichuris trichiura* have a unique football shape and are approximately 22–50 μm in length (Fig. 14.4).

TOXOCARA CANIS AND TOXOCARA CATI

The adult worms of *Toxocara canis* and *Toxocara cati* live in the lumen of the small intestine of dogs and cats, respectively. In these animals, toxocarial infection is cosmopolitan and occurs wherever there are significant dog and cat populations. The life cycle of these worms in dogs and cats is similar to that for *Ascaris lumbricoides*. *Toxocara canis* and *Toxocara cati* cannot complete their life cycle in humans. However, if humans are infected by the ingestion of an embryonated egg, the larva

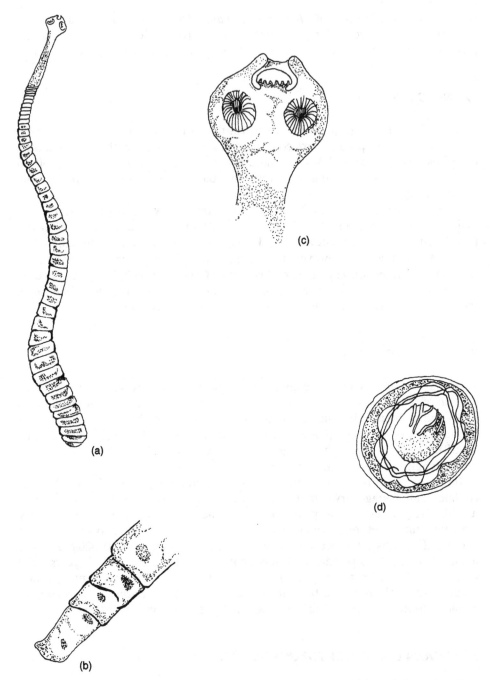

Figure 14.9 Hymenolepis nana. Hymenolepis nana: *adult tapeworm (a), posterior segments or proglottis (b), scolex (c), and ovum or egg (d). The posterior proglottis of the tapeworm is sterile and not capable of producing viable eggs.*

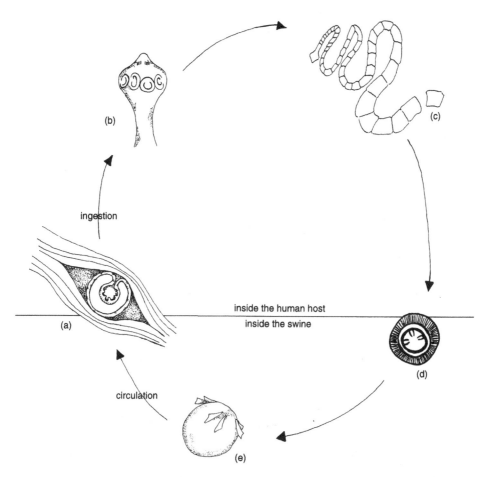

Figure 14.10 *Life cycle of* Taenia solium. *The life cycle of* Taenia solium *or the pork tapeworm begins when uncooked or improperly prepared pork containing cysticerci are ingested (a). Within the digestive tract of the human host small tapeworms emerge from the cysticerci and attach to the wall of the small intestine by their scolex (b). Here, the tapeworms grow in size and mature (c). Fertile ova or eggs are produced in the proglottids and released from the human host in fecal waste (d). When eggs are ingested by swine, onchospheres (e) hatch from the eggs and penetrate the intestinal wall of the swine. The onchospheres travel through the circulatory system and eventually become embedded in muscular tissue and form cysticerci.*

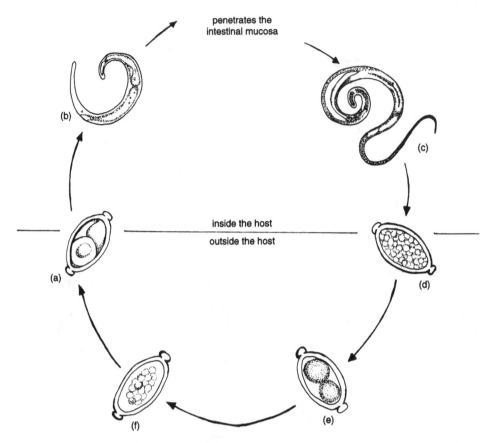

penetrates the
intestinal mucosa

(b)

(c)

inside the host

outside the host

(a)

(d)

(f)

(e)

Figure 14.11 *Life cycle of* Trichuris trichiura. *The life cycle of* Trichuris trichiura *or the whipworm begins when embryonated ova or eggs are ingested (a). Immature worms or larvae hatch from the eggs in the intestinal tract (b). The larvae penetrate the wall of the intestinal tract for a short period of time and then return to the lumen of the large intestine, where they attach to the intestinal mucosa and mature (c). The adult worms release fertilized eggs that leave the host in fecal waste (d). Outside the host, the eggs undergo development (e–f), until embryonated eggs are formed.*

can hatch from the egg and migrate from the intestine to produce a serious syndrome called visceral larval migrans. This disorder is serious and is caused by the circulating third larval stage migrating and damaging the liver, lungs, brain, or other organs. The eggs are 75–80 μm in length and contain a resistant shell. The eggs are similar in appearance.

Unlike protozoan cysts, it is easier to identify the ova of parasitic worms (Fig. 14.12) from stool or sputum samples. For intestinal parasites, a variety of drug treatments have been developed. More complicated procedures such as surgery or chemotherapy are often required to treat larval stages of helmintic parasites that may have migrated to tissues such as muscle and brain.

To identify parasitic eggs within wastewater or sludge, relatively large volumes of wastewater or sludge must be used. The eggs within the wastewater or sludge must be concentrated and then examined microscopically. Techniques for the isola-

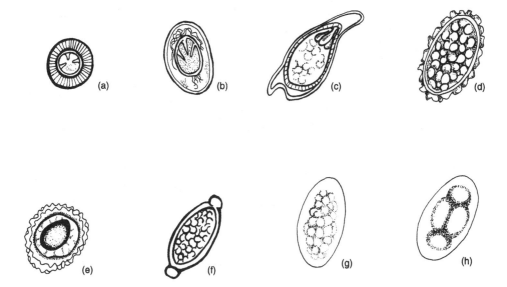

Figure 14.12 *Helminth ova. Because of the differences in shape, size, and structure, helminth ova or eggs often can be easily identified. Examples of ova include* Taenia solium *(a)*, Hymenolepis nana *(b)*, Schistosoma *sp. (c)*, Ascaris lumbricoides *(d)*, Toxocara canis *(e)*, Trichuris trichiura *(f)*, Necator americanus *(g)*, and Strongyloides stercoralis *(h)*.

tion of parasitic stages from wastewater or sludge are basically the same as those used for clinical parasitological detection of eggs in humans or animals.

There are two basic categories or techniques for sampling and identifying parasitic stages. These techniques are floatation analysis and sedimentation analysis. In these techniques helminth eggs are separated from debris through their floatation in higher-density solutions, namely, solutions having a specific gravity ≥1.18. However, some eggs of tapeworms and nematodes are too dense to float. These eggs must be removed from the debris through sedimentation analysis. The use of floatation analysis and sedimentation analysis can be accompanied by the use of various stains.

Part IV

Ectoparasites and Rodents

15

Ectoparasites

There are four ectoparasites that are carried by rats that are of concern to wastewater personnel. These parasites are a nuisance, cause infestation, and carry human diseases. These parasites are fleas, lice, mites, and ticks (Fig. 15.1).

Fleas are small, laterally compressed, wingless insects. Fleas temporarily infest birds and mammals for feeding purposes. Although humans are the principal host of two species of fleas, humans occasionally are infested with the fleas of other mammals (Table 15.1).

Fleas are vectors for several diseases, and their bites may result in dermatitis. Rodent species of fleas transmit bubonic plague and endemic (murine) typhus fever (Table 15.2). Bubonic plague or Black Death is caused by the bacterium *Yersinia pestis* and is primarily a disease of wild rodents. Humans typically are infected with *Yersinia pestis* from fleas that transmit the pathogen from rat to rat. The disease in humans is due chiefly to rats associated with human habitations, such as the Norway rat, *Rattus norvegicus*, and black rat, *Rattus rattus*. On the death of the rat, the infected fleas seek new hosts, either other rats or humans. Bubonic plague also may be acquired by direct contact with rodents. The disease endemic (murine) typhus fever is transmitted from rats to rats and from rats to humans by fleas. The infectious agent for endemic typhus fever, *Rickettsia typhi*, does not seriously harm the flea.

The bacterium *Yersinia pestis* is transmitted from infected rat to noninfected rat by the bite of an infected flea seeking blood for food. Once the flea is infected, the bacterium lives in the flea's stomach. As the bacterium reproduces in the stomach, it becomes difficult for the flea to digest the ingested blood. Therefore, on biting another host (rats or human) for food, the flea vomits into the bite of its host, causing the host to become infected with *Yersinia pestis*.

Although rats die from infection from *Yersinia pestis*, the bacterium may lie dormant in dead rats and may be transmitted to rats or humans by fleas. An indi-

Wastewater Pathogens, by Michael H. Gerardi and Mel C. Zimmerman
ISBN 0-471-20692-X   Copyright © 2005 John Wiley & Sons, Inc.

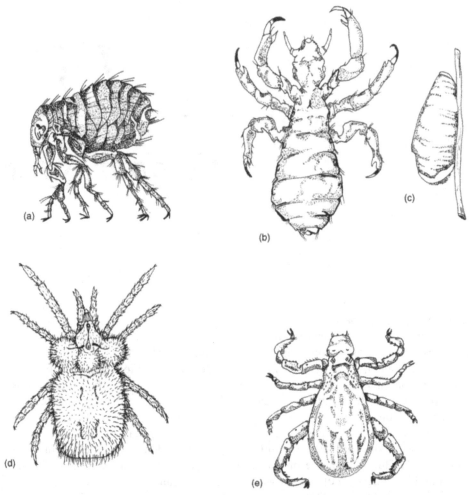

Figure 15.1 Ectoparasites. Several common ectoparasites of humans and animals include the flea (a), the louse (b), the mite (d), and the tick (e). The egg of the louse is cemented to a hair shaft (c) and is known as a nit.

TABLE 15.1   Fleas Associated with Humans or Human Disease

Flea	Common Name	Significance
Ctenocephalides canis	Dog flea	Infests dogs, may attack humans
Ctenocephalides felis	Cat flea	Infests cats, may attack humans
Leptopsylla segnis	Mouse flea	Common parasite of the house mouse, rat, and other small rodents
Nosopsyllus fasciatus	Rat flea	Principal host is Norway rat; infests other rats and rodents
Pulex irritans	Human flea	Most common flea found on humans; also infests dogs, mice, and rats

vidual can become infected with *Yersinia pestis* by being bitten by an infected flea, by direct contact with bodily fluids from an infected rat, and by inhalation of contaminated aerosols or particles. The incubation period for bubonic plague is 4–6 days.

TABLE 15.2 Significant Diseases Transmitted by Fleas

Disease	Infectious Agent
Bubonic plague	*Yersinia pestis*
Dog tapeworm	*Dipylidium caninum*
Dwarf tapeworm	*Hymenolepis nana*
Endemic typhus	*Rickettsia typhi*
Tularemia	*Pasteurella tularensis*

TABLE 15.3 Human Diseases Transmitted by Ticks

Infectious Agents	Disease
Rickettsial	American spotted fever
	Q-fever
Viral	Colorado tick fever
	Encephalitis
	Lyme disease
Bacterial	Relapsing fever
	Tularemia

Lice are small, dorsoventrally (top to bottom) flattened, wingless insects. There are two types of lice, biting lice and sucking lice. Lice are ectoparasites of birds and mammals, and an infestation of lice is referred to as pediculosis.

Lice are cosmopolitan in geographic distribution. They have played a critical role in the spread of epidemics of endemic typhus fever and relapsing fever. The causative agent for relapsing fever is the spirochete *Borrelia recurrentis* and is transmitted by the body louse *Pediculus humanus* variety *corporis*. The spirochetes are ingested by the louse with the blood of the infected host, multiply in the louse, and are distributed throughout the body of the louse in least than one week. An individual becomes infected with the spirochete when the crushed body of the louse comes in contact with the bite wound or abraded skin.

There are three lice that cause pediculosis in humans. These lice are *Pediculus humanus* variety *capitis* (head louse), *Pediculus humanus* variety *corporis* (body louse), and *Phthirus pubis* (crab louse).

Mites and ticks, along with scorpions and spiders, are members of the class Arachnida. They have four pairs of legs that are attached to the ventral surface of the body. Mites and ticks are very similar in appearance and structure. The term "mite" is used to describe members of the order Acarina other than ticks. Compared with ticks, mites are smaller in size and lack a leathery "skin." The larvae of mites are known as harvest mites, red bugs, and chiggers. There are free-living and parasitic mites. Parasitic mites are ectoparasites of the skin, mucous membranes, and feathers. Some mites are plant parasites.

Ticks are bloodsucking ectoparasites of mammals and birds. Ticks are larger in size than mites and have a leathery "skin." There are approximately 300 species of ticks, and most species are capable of biting humans. A few species of ticks transmit human disease (Table 15.3).

There are two types of ticks, soft ticks and hard ticks. Soft ticks are primarily ectoparasites of birds. Hard ticks possess a horny shield (scutum) that covers the dorsal surface of the tick. Hard ticks, especially species of the genus *Ixodes*, are injurious to humans.

16

Rodents and Rodent Control

Rodents, especially rats, within sewer systems and their control are of concern to wastewater personnel for several reasons. Rats cause structural and material damage. They are a public nuisance. Their nesting sites provide for the breeding of large numbers of fleas and mites, and, most importantly, they are carriers of disease. Occasionally, rats bite people.

There are over 50 species of rats, and two of these species are commonly found in sanitary sewer systems. These rats are the Norway rat (*Rattus norvegicus*) and the roof rat (*Rattus rattus*). The Norway rat is known also as the brown rat, wharf rat, sewer rat, water rat, on gray rat. The roof rat is known also as the black rat or house rat. There are significant differences in the appearance and feeding habits of these rats (Tables 16.1 and 16.2). Norway rats usually build nests in burrows under buildings, low bushes, ground cover, woodpiles, and garbage dumps. Because the roof rat is a better climber than the Norway rat, the roof rat is likely to build nests in walls, attics, and trees.

The number of rats and their behavior change throughout the year (Table 16.3). The changes in numbers and behavior are due to different environment conditions that exist each season.

The Norway rat and roof rat are commensal in nature and are not protected by law. Both rats are found throughout the United States. The Norway rat is found in all the contiguous states, whereas the roof rat is found on the West Coast and southeastern and Gulf Coast states, especially in seaports of these states (Table 16.4). Rats are not nocturnal animals, but they may be more active at night as a defensive behavior to protect themselves from predators, if predators are active in their environment.

Both rats possess significant physical abilities. They are excellent jumpers, climbers, and swimmers. They can jump 3′ vertically and 4′ horizontally and can fall

Wastewater Pathogens, by Michael H. Gerardi and Mel C. Zimmerman
ISBN 0-471-20692-X   Copyright © 2005 John Wiley & Sons, Inc.

TABLE 16.1   Appearance and Feeding Habits of Sewer Rats

Feature	Norway Rat	Roof Rat
Weight	~1 pound	~0.5 pound
Fur—back	Brownish or reddish-gray	Black to slate
Fur—belly	Whitish-gray	White
Tail—color	Bicolor-like body	Black covered with white scales
Tail—length	Shorter than head and body combined	Longer than head and body combined
Feeding habit	Omnivorous	Primary vegetarian
Nesting sites	Usually ground level or beneath the ground	Usually above ground level

TABLE 16.2   Comparison of Structural Features between the Norway Rat and the Roof Rat

Feature	Norway Rat	Roof Rat
Nose	Blunt	Pointed
Eyes	Small	Large
Ears	Small	Large
Body	Heavy and thick	Light and slender

TABLE 16.3   Seasonal Changes in Numbers and Behavior of Rats

Season	Population Number	Significant Environmental Condition
Spring	Annual breeding cycle starts in early spring; by late spring rats are abundant.	Increase in vegetative growth provides additional food, shelter, and breeding sites.
Summer	Populations are high and are commonly observed; additional peak breeding begins in late summer.	Vegetative growth is very abundant, permitting large rat populations.
Autumn	Rats are abundant; rats begin to seek winter nesting sites.	Vegetative decreases as well as food and shelter.
Winter	Populations are low; breeding is minimal; many rats do not survive cold temperatures and lack of food; rats seek warm shelter.	Vegetative is very low; food continues to decrease.

TABLE 16.4   Range of the Roof Rat

Alabama	Mississippi
Arkansas	North Carolina
California	Oregon
Delaware	South Carolina
Florida	Tennessee
Georgia	Texas
Louisiana	Virginia
Maryland	Washington

TABLE 16.5 Examples of Defects in Sanitary Sewers That Permit the Entrance of Rats

Improperly backfilled trenches allowing separation of sections of sewer mains
Insufficient thickness of manhole bottoms allowing rats to chew through manholes
Invasion of tree roots
Mains laid at an excessive grade allowing separation of sections of sewer mains
Poor connections between sections of sewer mains

Figure 16.1 Trichinosis, cysticerci of Taenia solium *in muscle tissue.*

50′ without serious injury. They can climb vertical surfaces with a rough exterior and can climb the outside and inside of vertical pipes. They can swim 0.5 miles in open water and against the water current. Also, they are capable of diving through plumbing traps and can swim underwater for 30 seconds at a time.

Rats gain entrance to the sanitary sewer system through several avenues (Table 16.5). Rats enter the sewer system through faulty joints or where tree roots have invaded sewer laterals and mains. Rats also enter the sewer system at manholes, catch basins, broken pipes, and broken drains.

Rats can enter the sewer through openings larger than a 0.5″ square. Because the incisors of sewer rats grow approximately 5″ per year, the rats must chew continuously. Therefore, they are capable of gnawing through aluminum sheeting, cinder block, fresh concrete, and lead sheeting. These physical abilities make "rodent proofing" of sanitary sewer systems labor intense and time consuming.

There are numerous health concerns related to the nesting and migration of rats in sewer systems. These concerns include the presence of ectoparasites and the source of disease. Ectoparasites include fleas, lice, mites, and ticks. Diseases associated with rats and their ectoparasites include bubonic plague (*Yersinia pestis*) carried by fleas, murine typhus fever (*Rickettsia mooseri*), rat-bite fever (*Streptobacillus moniliformis*), salmonellosis (*Salmonella* spp.) and trichinosis (*Taenia* spp.) (Fig. 16.1). Perhaps the two most important diseases of concern to wastewater personnel that are carried by rats are leptosporosis (*Leptospira* spp.) and dwarf tapeworm (*Hymenolepis nana*).

The normal home range for rats is less than 500 feet in diameter. However, rats may move great distances when disturbed. There are several reasons why rats move in large numbers into sewer systems. These reasons include a change in the carrying capacity of their environment and the favorable habitat offered by the sewer system. Demolition, redevelopment, and highway construction often are associated with the dispersal of rats into sewer systems. Reproduction in the spring and summer may result in an increase in population size that exceeds the ability of the environment (carrying capacity) to support the rapidly growing population. Overpopulation of the environment results in the death and dispersal of many rats.

Rodents quickly disperse or migrate to a sewer system. Conditions within the sewer system provide an ideal habitat for rodents. Water and food are readily available, and the sewer system offers numerous nesting sites and warmth during winter. Food is available in the sewers as floating food scraps, fecal material, and insects, such as cockroaches. The use of garbage disposals provides rats with a large quantity of undigested food. Therefore, relatively flat sewers that convey large quantities of garbage disposal wastes may provide rats with very favorable habitats. Rats also are protected from predators when they travel in the sewer system, and the sewer system allows the rats to move nonrandomly over relatively long distances.

Eradicating rats from sewer systems is difficult. Eradication requires a community effort. Rats are found in not only sanitary sewers but also storm sewers, woodpiles, lots with high vegetation, garbage, and sites with animal wastes. Rats often are present in sewers in relatively large numbers. The female is capable of mating more than once in a year and having 4 to 7 litters per year. The female may have more than 80 offspring each year. Population densities for rats vary greatly. Rats may be present at 100 to 200 per city block. Most rats usually live less than six months.

There are many openings to sanitary sewers that make it difficult to control the movement of rats. Rodents enter and leave the sewer system through broken mains and laterals, catch basins and combined sewers, and abandoned and unsealed laterals.

Openings to sanitary sewers can be located through field inspections, dye testing, camera inspection (televising) and smoke testing. However, smoke is an irritant and may cause rats to quickly leave their nests in large numbers and appear above ground on streets, lawns, and sidewalks. The sudden movement of large numbers of rats may alarm community members. Therefore, careful consideration should be given to the location and time of smoke testing as well as advance notification of smoke testing to community members.

Repairing broken sewers and disconnecting catch basins can seal openings to sanitary sewers. Abandoned laterals should be sealed with concrete, and open and abandoned basement drains should be plugged or capped. Repairs to sanitary and storm fixtures should be periodically inspected, because rats may reopen sealings and chew through freshly poured concrete.

Additional measures that may be taken to reduce rodent numbers and activity in sanitary sewers include proper sanitation measures and elimination of shelters (Table 16.6). These measures reduce the availability of food, water, and shelter for rats. Where rodents are active food should be stored in rodent-proof containers, garbage should be disposed of properly, and vegetation should be kept low.

TABLE 16.6 Rodent Control: Proper Sanitation Measures and Eliminating Shelters

Rodent Control	Examples
Proper sanitation	Clean up fallen fruits and nuts from trees. Don't scatter food for wildlife, such as birds and squirrels. Garbage—stored in rodent-proof containers. Garbage—use tight fitting lids for garbage cans. Provide for rapid decomposition of compost piles. Prune seed pods from plants. Store garden/lawn seeds and pet foods in rodent-proof cans.
Eliminating shelters	Cover foundation vents. Prune shrubs away from the ground. Repair cracks and small holes in foundations. Repair windows and screens. Seal openings where pipes and wires enter buildings.

Rats also may be controlled through the use of a chemical or poison. Rat poisons are known as rodenticides, and numerous rodenticides are registered for use. There are acute (single dose) toxicants that contain arsenic or phosphorus, single-dose anticoagulants such as brodifacoum and bromadiolone, and multiple-dose anticoagulants such as diphacionone and warfarin. Anticoagulants cause internal bleeding and death.

Most multiple-dose anticoagulants are popular and ready to use, and they provide a margin of safety that single-dose anticoagulants do not. Multiple-dose anticoagulants require several feedings before a rat dies of internal bleeding. This multiple feeding requirement provides a safety factor for children and pets that may accidentally eat the poison. Therefore, children and pets are able to receive necessary medical attention when symptoms of poisoning occur.

Anticoagulants inhibit the production of prothrombin, the agent responsible for blood clotting. Therefore, when capillaries are damaged by the ingestion of an anticoagulant, internal bleeding occurs.

Anticoagulants are relatively inexpensive and easy to use. They are available as 2-ounce pellets or 1-pound cakes (rat cakes, bars, or blocks). If rats become tolerant of one anticoagulant, another anticoagulant may be used.

Rat cakes should be secured in manholes so they do not float away or be carried off by rats. Rat cakes often are suspended by wire in manholes approximately 3″ over the workbench. Sanitary manholes may be baited with one or more rat cakes, depending on the severity of rodent infestation and feeding.

Wherever a rodent infestation is found, all sanitary manholes within a three-block radius of the center of infestation should be baited with rat cakes. Each baited manhole should be inspected periodically, often daily, to determine the degree of rodent feeding and the area of infestation. Continuous feeding from 1 to 2 weeks should provide control. The radius of the baited manholes should be increased or decreased according to the degree of rodent feeding at the perimeter of the baited area. Rats typically die in out-of-the-way places, such as burrows.

Around wastewater treatment plants, rodenticides can be made available to rats in well-constructed bait stations or boxes (Fig. 16.2). The boxes are tamper resistant and usually have two openings that are approximately 2 inches in diameter. Each entrance is bileveled, and the box contains internal baffles. These features prevent

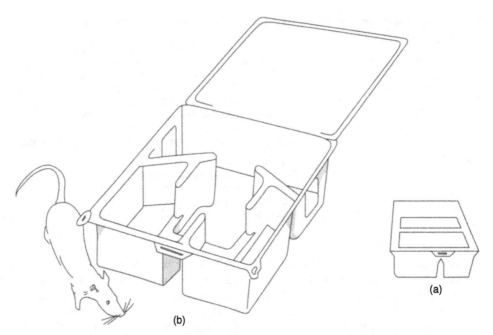

Figure 16.2 *A bait station for the control of rats. A bait station for the control of rats is shown closed (a) and opened (b). In the closed position the station is anchored to the ground from the inside and the openings for rats are too small to permit the entrance of cats and dogs. Also, the internal structure of the station contains internal baffles that prevent children from reaching into the station and removing the bait.*

a small child from reaching into the bait box and a pet from entering the bait box. The top and bottom of the box are secured with plastic straps. The bait or 2-ounce pellets of rodenticide are secured in the box, and the box is anchored in the ground.

When rats move they usually place their whiskers and body hair in contact with a wall or other object. Therefore, bait boxes should be placed next to walls. The location of bait boxes next to walls doesn't interfere with the natural movement of rats, and rats will enter the bait box.

HYMENOLEPIS NANA

The dwarf tapeworm, *Hymenolepis nana*, is cosmopolitan in geographic distribution and is commonly found throughout the United States. It is the most common tapeworm in humans in southeastern United States.

The dwarf tapeworm is relatively small in size. Adult worms range from 30 to 40 mm in length and from 0.5 to 1.0 mm in width. As many as 200 segments or proglottids (Fig. 16.3) and the head or scolex (Fig. 16.4) make up the worm. The tapeworm attaches to the intestinal wall by the small globular scolex. Mature proglottids produce eggs or ova.

Eggs are oval or globular in shape and are enclosed in two membranes. There is considerable variation in the size of the ova (30 to 60 µm). Each mature proglottid is capable of producing 80 to 180 ova. The ova have a low resistance to cold, heat, and desiccation and cannot survive long outside the host.

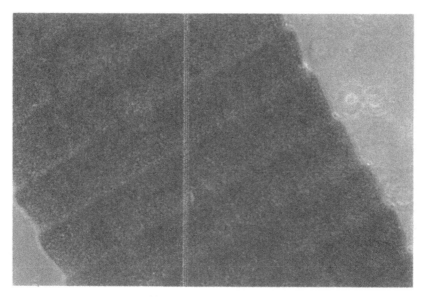

Figure 16.3 *Proglottids of* Hymenolepis nana.

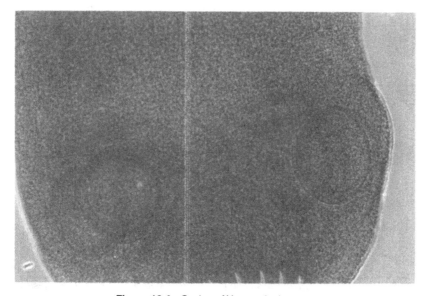

Figure 16.4 *Scolex of* Hymenolepis nana.

Hymenolepis nana is a common parasite of humans, mice, and rats. When gravid (ova laden) proglottids rupture within the intestines, ova are released. The released ova are infective and pass from the host in the feces.

When a new host ingests an ovum, the oncosphere is liberated in the small intestine (Fig. 16.5). The oncosphere penetrates a folding or villus of the small intestine. Over the next four days the oncosphere becomes a cercocystis or larva. The larva

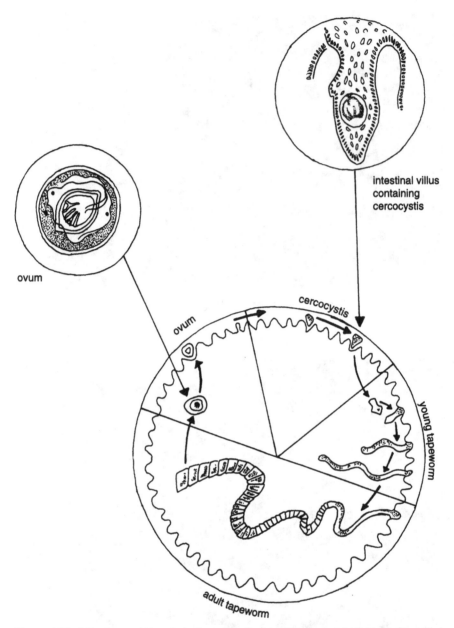

ovum

intestinal villus
containing
cercocystis

Figure 16.5 *Life stages of Hymenolepis nana in the intestinal tract of its host. The ingested embry-onated ovum of Hymenolepis nana penetrates into the intestinal wall. From the ovum the onchosphere hatches and develops into a cercocystis in an intestinal villus. A young tapeworm emerges from the villus and attaches to the intestinal wall by its scolex. Here, the young tapeworm matures to an adult worm.*

eventually breaks free of the villus and moves into the lumen of the small intestine. Here, it attaches to the epithelium of the intestine. In 10–12 days the larva becomes an adult worm. Approximately 30 days are required from the time of infection until ova appear in the feces of the host.

Part V

Aerosols, Foam, and Sludge

17

Aerosols and Foam

AEROSOLS

Wastewater treatment plants are a source of airborne pathogens. Dust and especially aerosols or bioaerosols carry pathogens and molecules such as allergens and toxins. Allergenic reactions from exposure to allergens and toxins are possible. Examples of viable bacteria that have been recovered from aerosols are numerous (Table 17.1).

Aerosols are water droplets that are <10μm in diameter and float on air currents. Aerosols are produced by bursting bubbles in aeration tanks and spraying or splashing in turbulent environments such as the headworks of treatment plants, sanitary sewers, sludge disposal sites, and trickling filters (Table 17.2). Airborne pathogens and aerosol inhalation represent a potential health risk to wastewater personnel and community members downwind of treatment plants, sludge disposal sites, and spray irrigation sites. Urban development has placed large populations in close proximity to treatment plants, and the need for more and more disposal sites for ever-increasing quantities of sludge have exposed rural communities to aerosols and pathogens.

Airborne pathogens consist of viruses, bacteria, and fungi. The infective stages of protozoans (cysts and oocysts) and helminths (eggs) are too heavy to be carried by aerosols.

Aeration is a major factor responsible for the production of viable bioaerosols. A bioaerosol is an airborne particle that contains a virus, bacterium, or fungus that is capable of forming a colony through replication or reproduction in controlled laboratory conditions.

Several factors affect the emission of bioaerosols. Aeration rates and type of aeration in activated sludge processes and updraft conditions in trickling filter processes are important factors. Bioaerosols increase in number with increasing

Wastewater Pathogens, by Michael H. Gerardi and Mel C. Zimmerman
ISBN 0-471-20692-X   Copyright © 2005 John Wiley & Sons, Inc.

TABLE 17.1 Genera of Viable Bacteria Recovered from Aerosols

Aerobacter	Proteus
Bacillus	Pseudomonas
Escherichia	Salmonella
Klebsiella	Shigella
Mycobacterium	Staphylococcus

TABLE 17.2 Wastewater Treatment Processes That Produce Bioaerosols

Activated sludge aeration tanks
Grease chambers
Grit chambers
Preaeration tanks
Primary clarifiers
Sanitary sewers
Secondary clarifiers
Sludge application sites using sprays or sprinklers
Trickling filters

Figure 17.1 Filamentous organism foam (Nocardioform) on an aeration tank.

TABLE 17.3 Commonly Occurring Foams at Activated Sludge Processes

Foam	Biological/Chemical	Texture and Color
Filamentous organism	Biological	Viscous and chocolate brown
Nutrient deficiency	Biological	Billowy white (young sludge)
		Greasy gray (old sludge)
Oil and grease	Chemical	Viscous dark brown
		Viscous black
Surfactant	Chemical	Billowy white
Zoogloeal growth	Biological	Billowy white

aeration rates and increasing flow in trickling filter updrafts. Turbulence within grit chambers also is responsible for the release of many bioaerosols from raw wastewater.

Emission of bioaerosols from aeration tanks and trickling filters increases with increasing wind speed. Increased wind action causes a larger number of bioaerosols to become airborne or remain airborne.

FOAM

Foam production and accumulation (Fig. 17.1) often occurs at activated sludge processes. Several types of foam may be produced (Table 17.3). These foams are biological or chemical in nature and consist of air or gas bubbles entrapped beneath a solid layer. Although foam is not produced by pathogens, it may contain pathogens. Therefore, the treatment and handling of foam should be performed with the use of proper hygiene measures and appropriate safety equipment.

18

Sludge

Sludge or wastewater residuals have been used as a fertilizer or soil conditioner in the United States for many years. Wastewater residual consists of solids collected during various stages of wastewater treatment. The residuals consist of grit, primary sludge, secondary sludge, and tertiary sludge. Grit is the material collected in the grit chamber. This material usually is placed in landfills. Primary sludge is produced through sedimentation in the primary clarifier. These clarifiers are located upstream of biological treatment systems such as the activated sludge process and the trickling filter process. Secondary sludge consists of solids, mostly bacterial, that are produced in biological treatment systems and are removed from the systems through sedimentation in the secondary clarifier. Tertiary sludge is produced in advanced treatment processes, often through the use of chemicals, for example, the addition of lime to remove phosphorus from the biologically treated effluent.

Biosolids consist of primary sludge and secondary sludge that have undergone additional treatment such as aerobic or anaerobic digestion of the wastes within the sludge (Table 18.1). The digestion of sludge has many purposes including reduction in volatile content of the sludge and destruction or reduction in number of pathogens.

The use and disposal of sludge in the United States is regulated under 42 CFR parts 257, 403, 503—Standards for the Use or Disposal of Sewage Sludge Final Rules. The U.S. EPA is the lead agency that has regulatory responsibilities for wastewater treatment and sludge disposal.

There are two separate pathogen reduction requirements for sewage sludge—Class A and Class B. The goal of Class A requirements is to reduce pathogen levels to below detection limits. The goal of Class B requirements is to reduce the level of pathogens to concentrations that are unlikely to pose a health risk to the public and the environment. However, there are site restrictions for land application of Class

Wastewater Pathogens, by Michael H. Gerardi and Mel C. Zimmerman
ISBN 0-471-20692-X   Copyright © 2005 John Wiley & Sons, Inc.

TABLE 18.1 Sludge Treatment Methods for the Destruction of Parasites and Pathogens

Methods	Impact on Parasites and Pathogens
Aerobic digestion	Mesophilic temperatures ineffective for many parasitic helminths
	Reduction of enteric bacteria and viruses at 40°C
	Thermophilic temperatures effective on all parasites and pathogens
Anaerobic digestion	1–3 log reduction in pathogenic bacteria
	3 log reduction in protozoan cysts
	<50% destruction of helminth ova at mesophilic temperatures
	Many viruses survive mesophilic and thermophilic temperatures
Air drying	Does not appear reliable for complete inactivation of parasites and pathogens.
Composting	Effective parasite and pathogen destruction at 55°C for 3–15 days
Dual digestion	Parasite and pathogen destruction greater than those achieved by thermophilic aerobic digestion
Lime stabilization	Effective in destroying pathogenic bacteria and viruses
	Moderately effective in destroying most parasitic helminths
Pasteurization	Effectively destroys parasites and most pathogenic bacteria and viruses
Sludge lagooning	Parasite and pathogen destruction highly variable; efficiency affected by local temperature and retention time

B sludge. Crop harvesting, animal grazing, and public contact are limited in an effort to allow environmental factors to further reduce pathogen levels.

Many pathogens that enter wastewater treatment facilities are removed from the waste stream and concentrated in sludge. Two basic processes of treatment are used to reduce pathogen levels in sludge. These processes consist of those that significantly reduce pathogens (PSRP) and those that further reduce pathogens (PFRP).

PROCESSES TO SIGNIFICANTLY REDUCE PATHOGENS

Processes to significantly reduce pathogens include: aerobic digestion (60 days at 15°C to 40 days at 20°C), anaerobic digestion (60 days at 20°C to 15 days at 35–55°C), lime stabilization (pH ≥ 12 after 2-hour contact time), and mesophilic composting (minimum: 40°C for ≥5 days). Most wastewater treatment plants are equipped with processes that qualify as PSRP. However, there are restrictions with regard to crop production, for example, no food crop should be grown within 18 months after sludge application, animal grazing (no grazing for at least 1 month by animals that provide products that are consumed by humans), and public access to the sludge disposal site (controlled access for at least 12 months).

PROCESSES TO FURTHER REDUCE PATHOGENS

Processes to further reduce pathogens include heat treatment (at 180°C for 30 minutes), irradiation, high-temperature composting, thermophilic aerobic digestion (10 days at 55–60°C), and heat drying. Some PFRP methods such as high-temperature composting, thermophilic aerobic digestion, and heat drying do not require prior PSRP treatment. However, PFRP processes are required if edible crops are exposed to sludge.

AEROBIC DIGESTION

Aerobic digestion is used by many small communities to process sludge. Detention time and temperature appear to influence the efficiency of these operations in destroying pathogens. Mesophilic aerobic digestion appears to be ineffective in destroying parasitic helminths. However, thermophilic aerobic digestion ($\geq 45°C$) appears to be more effective in destroying parasitic helminths.

ANAEROBIC DIGESTION

Most of the sludge in the United States is treated by anaerobic digestion. Detention time, temperature, and sludge feeding protocols are the major factors affecting pathogen survival during anaerobic digestion of sludge. Generally, the draw/fill mode of operation of digesters (the digested sludge is withdrawn before feeding the digester with sludge) achieves greater pathogen reduction than the fill/draw mode of operation. Critical temperatures for the destruction of pathogens are 20–50°C for mesophilic digestion and 55–80°C for thermophilic digestion.

AMMONIFICATION

Ammonification is the addition of ammonia (NH_3) or ammonium sulfate [$(NH_4)_2SO_4$] to sludge. Although this method is effective in destroying large numbers and a large variety of pathogens, the method is expensive.

AIR DRYING

During air drying sludge is placed on sand beds, with water removal achieved through drainage and evaporation. Sludge drying to 95% solids appears to be effective in reducing the bacterial pathogens. However, sludge drying is not effective against helminth ova and viruses unless heat ($\geq 55°C$) is applied.

CHEMICAL INACTIVATION

Chemical inactivation is the addition of lime slurry to sludge to achieve a pH of ≥ 12. Several modifications of this process exist, with best results achieved when pH is maintained >12 for at least 2 hours.

COMPOSTING

Composting involves mixing sludge with organic fillers such as wood chips. Temperature is the major factor controlling pathogen destruction. However, because of the heterogeneous makeup of the compost material, an adequate temperature for pathogen destruction usually is hard to maintain. Maintenance of a temperature of

55°C for 3–15 days appears to be critical for pathogen destruction. However, composting does present a health hazard for composting operators because of the growth of pathogenic fungi such as *Aspergillus fumigatus* during the composting operation. The use of proper hygiene measures and appropriate safety equipment should be practiced by composting operators to prevent the inhalation of fungal spores.

DUAL DIGESTION SYSTEMS

Dual digestion systems (DDS) consist of two steps: (1) utilization of a covered aerobic digester using pure oxygen and usually a 1-day detention time with heat (often ≥55°C) and (2) utilization of an anaerobic digester, usually with a 8-day detention time. Most destruction of pathogens occurs in the aerobic digester.

HEAT TREATMENT

Heat treatment involves heating sludge under pressure to 260°C for at least 30 minutes. Pathogens are effectively destroyed under these conditions.

IRRADIATION

Under irradiation sludge is subjection to radiation released by cesium 137 or cobalt 60 or high-energy electron beams. Although highly successful in destroying pathogens, this method is expensive and the perception of generating radioactive sludge is a problem.

MICROWAVE TREATMENT

This treatment method has been proposed, and various prototypes have been tested. Results of pathogen destruction vary with microwave treatment.

OZONICS

Ozonics is another expensive treatment method. It involves the addition of ozone (O_3) at a high concentration (200 ppm), low pH (2.5–3.5), and high pressure (60 psi). Destruction of parasitic helminths is variable with this method.

PASTEURIZATION

Pasteurization is the application of high temperatures (>70°C) for 30 minutes. Several prototypes for pasteurization units have been developed to save money by lowering temperature and exposing sludge pathogens for a longer period of time.

SLUDGE LAGOONING

Pathogen inactivation in sludge lagoons depends on local weather conditions, detention time, and type of pathogens in the lagoon. Storage time for 1-log reduction of pathogens is usually 1, 2, and 6 months for bacteria, viruses, and parasitic helminth eggs, respectively, at temperatures >20°C. Longer detention times are needed at colder temperatures.

EPIDEMIOLOGICAL SIGNIFICANCE OF PARASITES AND PATHOGENS IN SLUDGE

Humans and animals can be exposed to parasites and pathogens through direct or indirect contact with sludge. Direct contact may result from walking in sludge-amended soil, handling crops grown on sludge-amended soil, or inhaling airborne pathogens generated at sludge application sites. Indirect contact may result from consumption of contaminated crops or products obtained from animals that grazed on sludge-amended pastures.

Epidemiological studies should be designed to examine the transmission of parasites and pathogens via sludge-amended soils. These studies also should be designed to monitor parasite and pathogen transmission over extended periods of time.

Part VI

Disease Transmission and the Body's Defenses

19

Disease Transmission

For a pathogen to cause disease (damage) in a host, six steps of pathogenesis must be satisfied (Table 19.1). In the first step, ***transmission***, the pathogen must leave an infected individual and be transported to a noninfected individual. This may be accomplished through wastewater, sludge, bioaerosol, foam, vectors, and contaminated materials.

Disease transmission often is described as occurring by contact, vehicle, or vector. Contact may be direct (person to person) or indirect (through clothing or eating utensils). Typical vehicles for disease transmission are food, air, and water. Food may be a vehicle for disease transmission if it is contaminated, improperly cooked, poorly refrigerated, or prepared under unsanitary conditions. Air is a vehicle for disease transmission when pathogens are present in bioaerosols, droplets, dust, mist, sprays, or mist. Contaminated water also is a vehicle for disease transmission

There are numerous carriers or vectors of human diseases. Common vectors include cockroaches, fleas, houseflies, lice, mites, mosquitoes, rats, and ticks. For example, the hairs on a housefly's body or the legs of a cockroach carry millions of pathogens. Therefore, controlling populations of vectors helps to reduce the risk of disease transmission.

In the second step of pathogenesis, ***entry***, the pathogen must enter a noninfected individual. This is achieved through a portal of entry. Pathogens may enter an individual by crossing the skin and mucous membranes, especially those of the gastrointestinal tract and respiratory tract. Entrance through the gastrointestinal tract (ingestion) is the most common route and is usually due to poor hygiene. Ingestion may occur through contaminated food, water, cigarettes and smokeless tobacco, splashes, or swallowing bioaerosols through the oropharynx (Fig. 19.1).

The pharynx, or "throat," forms a skeletal muscular tube. It is divided into several continuous chambers. These chambers are the isthmus of the fauces or the initial

Wastewater Pathogens, by Michael H. Gerardi and Mel C. Zimmerman
ISBN 0-471-20692-X   Copyright © 2005 John Wiley & Sons, Inc.

TABLE 19.1   Steps of Pathogenesis

Transmission
Entry
Adherence
Multiplication
Spread
Damage

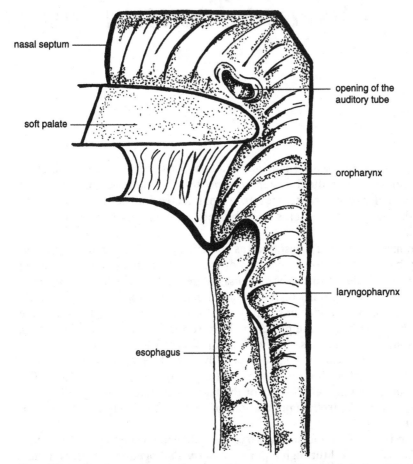

Figure 19.1 Oropharynx. Many pathogens, bioaerosols, and contaminated particles that enter the respiratory tract through the laryngopharynx are captured by mucus and cilia that line the respiratory tract. Once captured, the pathogens, bioaerosols, and contaminated particles are transported by ciliary action to the oropharynx, where they are swallowed into the digestive tract through the esophagus.

chamber, followed in turn by the oropharynx and the laryngopharynx. Muscles associated with the pharynx, for example, the constrictors, squeeze the pharynx in sequence, from top to bottom, initiating swallowing and peristalsis.

Many bioaerosols and dust particles entering the upper respiratory tract are captured by cilia that line the upper respiratory tract, and through the beating action of cilia they are moved into the oropharynx, where they are swallowed into the digestive tract.

The third step of pathogenesis is *adherence*. Once in the new host, the pathogen must find its target host cell and invade the cell. For example, hepatitis viruses must invade cells of the liver, whereas *Mycobacterium* spp. must invade cells of the respiratory tract. If the pathogen is successful in invading its target cell, the agent must replicate or reproduce (multiply). *Multiplication* is the fourth step in pathogenesis. During this step, the pathogenic agent may increase in numbers sufficient to cause disease (damage). The period of time that the pathogen takes to increase to sufficient numbers to cause disease is the incubation period.

If the pathogen spreads from the targeted (infected) area, it may damage cells close to the infected area or far from the infected area. *Spread* is the fifth step in pathogenesis. It may be accomplished directly by the increasing number of pathogen or indirectly by the host. For example, the pathogen may be carried and spread by white blood cells.

Disease or *damage* is the sixth step of pathogenesis. Disease is a change in the health of an individual. Disease may be associated with harsh symptoms such as the death of cells (gangrene) or mild symptoms such as a rash. Disease also may be associated with physiological changes such as the production of pathogenic toxins that interfere with bodily functions. For example, tetanus occurs with the production of bacterial toxins that interfere with the body's ability to transmit nervous impulses.

INFECTION

Although pathogens may gain entrance to the body through a portal of entry, this does not mean that the body will experience damage or disease. There are several criteria that must be satisfied for disease to occur. First, an infective dose of the pathogen must enter the body. The infective dose may be relatively small, for example, four–six cysts for *Giardia lamblia*, or relatively high, for example, several hundred bacteria for *Salmonella typhi*.

Second, the invading pathogens must be virulent, that is, they must be capable of causing disease. Dead pathogens cannot cause disease, and weakened or damage pathogens cannot or are highly unlikely to cause disease. Finally, the pathogens must overcome the body's defense mechanisms. These mechanisms consist of nonspecific and specific defenses against infection.

20

The Body's Defenses

Although infections from specific pathogenic agents are not common, wastewater personnel, especially during the first few years of employment, experience some increased symptoms of gastrointestinal and upper respiratory tract illness. For any illness or disease to occur, pathogens must overcome the body's defenses.

Wastewater personnel with a weak immune system continually become ill, whereas wastewater personnel with a strong immune system become ill less often. However, over time many wastewater personnel develop resistance to pathogens through their daily exposure to them. This occurs through "low-dose challenge" when the immune system develops improved immunity through repeated exposure to the same pathogens.

The body's defenses against pathogenic infection and disease occurrence can be placed into two groups, nonspecific and specific defenses. Nonspecific defenses (Table 20.1) operate regardless of the invading pathogen. Specific defenses (Table 20.2) operate against specific invading pathogens. Nonspecific defenses are the body's first line of defense. These defenses consist of physical barriers, phagocytes, antimicrobial compounds, and inflammatory response. Specific defenses are the body's second line of defense and include specific immune responses and immunization. The immune responses are specific because each response is "triggered" by a specific antigen (pathogen) and the triggered response is specific for that antigen.

Wastewater Pathogens, by Michael H. Gerardi and Mel C. Zimmerman
ISBN 0-471-20692-X   Copyright © 2005 John Wiley & Sons, Inc.

TABLE 20.1   Significant Nonspecific Defenses, the First Line of Defense

Defense	Component
Physical barriers	Skin, eyes and ears, respiratory system, digestive system, cardiovascular system, lymphatic system
Phagocytes	Granulocytes, agranulocytes

TABLE 20.2   Specific Defenses, the Second Line of Defense

Specific Immunity
Immunization

TABLE 20.3   Examples of Normal or Indigenous Bacteria of Various Body Sites

Body Site	Genera of Indigenous Bacteria
Skin	*Corynebacterium, Lactobacillus, Micrococcus, Propionibacterium, Staphylococcus*
Oral cavity	*Bacteroides, Fusobacterium, Streptococcus*
Gastrointestinal tract	*Bacteroides, Clostridium, Escherichia, Klebsiella, Lactobacillus, Streptococcus*
Upper respiratory tract	*Bacteroides, Staphylococcus, Streptococcus*

NONSPECIFIC DEFENSES

Physical Barrier: Skin

The skin (epidermis and dermis) is the first barrier. The skin forms a protective barrier that blocks the entry of pathogens. The skin also protects the body from pathogenic invasion through several mechanisms.

Acidic compounds in sebum—the oily substance secreted by the sebaceous glands in the skin—produce and maintain a pH range of 3 to 5 on the skin. In addition, lactic acid from lactobacilli that inhabit the skin contribute to a low pH, and many indigenous bacteria degrade secretions from oil glands that result in the release of free fatty acids that help to maintain a low pH. A low pH is inhibitory to some pathogens.

Many metabolic products of normal skin flora or bacteria (Table 20.3) inhibit the growth of some pathogens. Examples include the unsaturated fatty acids produced by *Staphylococcus epidermidis* and *Propionibacterium acnes*. These acids are highly toxic to Gram-negative bacteria.

Many indigenous bacteria also help to prevent infections by entrapping or inactivating pathogens. Many indigenous bacteria produce harsh growth conditions that inhibit or destroy pathogens. Harsh growth conditions are produced by competition for available nutrients, release of toxic compounds, and reduction in oxygen concentration. Indigenous bacteria that help to prevent pathogenic infections are numerous. The human body has approximately 10^{13} human cells and approximately 10^{14} associated nonhuman cells.

Physical Barrier: Eyes and Ears

The eyes have several protective external structures. These structures include eyelids, eyelashes, mucous membranes, and the cornea. The eyes also have a lacrimal gland that produces tears that flush foreign bodies from the eyes. The tears contain lysozyme, an enzyme that kills bacteria by destroying their cell wall. Lysozyme also is found in many bodily fluids including mucus, saliva, and sweat.

The ears also have protective features. The ear canal is lined with many small hairs and numerous ceruminous glands. The glands secrete cerumen or ear wax. The wax as well as the lining of hair helps to keep pathogens from entering the ear canal.

Physical Barrier: Respiratory System

The respiratory system consists of the upper respiratory tract and the lower respiratory tract. The upper respiratory tract includes the nasal cavity, pharynx, larynx, trachea, bronchi, and large bronchioles. The lower respiratory tract includes thin-walled bronchioles and alveoli, where gas exchange occurs. The entire respiratory system is lined with moist epithelium. The epithelium in the upper respiratory tract contains mucus-secreting cells and is covered with cilia.

Several mechanisms are active in protecting an individual from pathogenic infection through the respiratory system. Secreted mucus from the membranes in the upper respiratory tract traps pathogens and particles that may contain pathogens. The mucus contains lysozymes that degrade the cell walls of bacteria, and coughing and sneezing not only expose pathogens to mucus but also help to expel them.

Cilia in the upper respiratory tract beat toward the pharynx. This action is referred to as the mucociliary escalator, and it lifts the pathogens to the oropharynx, where they are spit out or swallowed.

Physical Barrier: Digestive Tract

The digestive tract consists of the mouth, pharynx, esophagus, stomach, and intestines and support organs such as the salivary glands, liver, and pancreas. The digestive system employs several mechanisms that attack invading pathogens.

Throughout the digestive tract mucin, a glycoprotein in mucus, coats many pathogens and keeps them from attaching to the inner surface of the digestive tract. Saliva also contains antibodies that destroy pathogens. Lysozyme also is produced and released in the digestive tract.

The stomach and small intestines destroy pathogens. The strong acidity of the stomach and bile acids and enzymes in the small intestine destroys many pathogens and inactivate many viruses. Acidity in the stomach is due in large part to the production of gastric acid or hydrochloric acid (HCl). The indigenous bacteria of the large intestine destroy and inactivated pathogens by surrounding them so they leave the host in feces. Inactivation or destruction of pathogenic agents in the lower intestinal tract continues with the maintenance of a low pH. Here, lactic acid ($CH_3CHOHCOOH$) and acetic acid (CH_3COOH) are produced through metabolic

TABLE 20.4   Major Formed Elements of the Blood

Formed Element	Function
Erythrocyte	Transport oxygen throughout the body
Leukocyte	Nonspecific and specific defenses
Platelet	Blood clotting component

fermentation. The indigenous bacteria of the intestinal tract also rob pathogens of essential nutrients. In the digestive tract lactoferrin and transferrin bind to iron and steal this essential growth nutrient from pathogens.

In addition to the digestive tract, lactoferrin and transferrin are found in other physical barriers. Lactoferrin is found in tears, bile, and nasopharyngeal, bronchial, and intestinal secretions. Transferrin is found in blood serum and the intercellular spaces of many tissues and organs.

Physical Barrier: Cardiovascular System

The cardiovascular system consists of the heart, blood vessels, and blood. The blood contains numerous formed elements (Table 20.4). These elements include (1) erythrocytes that contain hemoglobin and transport oxygen throughout the body, (2) leukocytes or white blood cells that contribute to nonspecific and specific defenses, and (3) platelets. Platelets are an important blood-clotting agent.

Physical Barrier: Lymphatic System

The lymphatic system consists of a network of vessels, lymph nodes, lymphatic tissue, and lymph fluid. Lymphatic tissue contains cells that phagocytize pathogenic agents. These cells are B lymphocytes (B cells) and T lymphocytes (T cells).

Phagocytes

Phagocytes or leukocytes (white blood cells) remove debris and pathogens from the body (Fig. 20.1). Leukocytes contribute to specific and nonspecific defenses. There are two groups of leukocytes—granulocytes and agranulocytes (Table 20.5). Granulocytes have a granular cytoplasm and irregularly shaped nuclei, whereas agranulocytes lack granules in the cytoplasm and have round nuclei.

There are three types of granulocytes. They are basophils, eosinophils, and neutrophils. Basophils migrate into tissues, where they are called mast cells. Inside the tissues the mast cells release histamines and heparin. Histamines initiate the inflammatory response, whereas heparin inhibits blood clotting.

Eosinophils are released in large numbers during allergic reactions. They act as phagocytes and may detoxify foreign compounds. Neutrophils are vigorous phagocytes. They protect the skin and mucous membranes from pathogen invasion.

There are two types of agranulocytes, monocytes and lymphocytes. Monocytes also migrate into tissues. Once inside tissues, monocytes are referred to as macrophages. As macrophages, they engulf pathogens and debris.

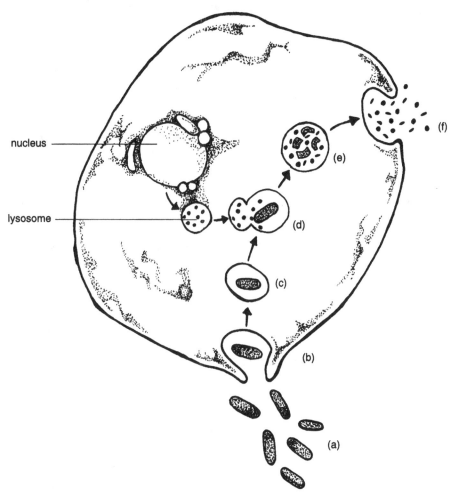

Figure 20.1 *Phagocytosis. The process of phagocytosis consists of several steps. When foreign bodies such as pathogens enter the human body (a), phagocytic cells within the immune system (b–c) gradually ingest the foreign bodies. Once ingested, the foreign bodies are digested by lysosomes (d–e). the wastes resulting from digestion are released from the phagocytic cells (f).*

TABLE 20.5   Groups of Leukocytes

Group	Leukocytes
Granulocytes	Basophils, eosinophils, neutrophils
Agranulocytes	Monocytes, lymphocyte B cells, lymphocyte T cells, lymphocyte NK cells

Lymphocytes (B cells, T cells, and NK cells) are carried in the blood and are found in large numbers in lymphoid tissues, where they contribute to specific immunity. Lymphoid tissue includes lymph nodes, spleen, thymus, and tonsils.

If pathogens are successful in infecting (colonizing) an individual, the individual may display no clinical symptoms (asymptomatic), mild symptoms (acute), or severe

TABLE 20.6 Disease Manifestations in Infected and Diseased Individuals

Manifestation	Clinical Symptoms
Asymptomatic	None
Acute	Mild
Chronic	Severe

symptoms (chronic) (Table 20.6). Regardless of the symptoms or lack of symptoms displayed by an individual, the bodily wastes or fluids of the infected individual do contain pathogens.

An example of an asymptomatic carrier of typhoid fever was Mary Mallon or Typhoid Mary. Mary Mallon was an Irish immigrant who lived and worked in New York City in the early 1900s. She was employed as a cook and was responsible for causing many cases of typhoid fever, a number of them fatal. Poor hygiene practices by Mary Mallon were responsible for the transmission of the typhoid bacterium, *Salmonella typhi*.

SPECIFIC DEFENSES

Significant specific defenses consist of immunity and immunization. These topics are reviewed in Chapter 25, *Immunization*.

Part VII

Removal, Inactivation, and Destruction of Pathogens

21

Removal, Inactivation, and Destruction of Pathogens

Pathogens that enter a wastewater treatment system may be freely dispersed, associated with the cells that they infect, or adsorbed to solids. There are several fates that can occur to these pathogens in a wastewater treatment system (Table 21.1). The pathogens may be removed, inactivated, destroyed, consumed by higher life-forms, or leave the treatment system. Pathogens that leave the treatment system in a viable or an inactivated condition are of concern.

REMOVAL

Pathogens may be removed from the wastewater by their adsorption to biological solids or inert solids. The biofilm within the sewer system adsorbs and filters out many pathogens. Some of these, such as *Leptospira interrogans*, grow in the biofilm inside manholes. Many pathogens are removed from the wastewater when solids on which they are adsorbed settle out in primary clarifiers. Some pathogens such as protozoan cysts and oocyts and helminth eggs settle out in large numbers in primary clarifiers because of their relatively high density.

Biological treatment processes such as activated sludge (suspended growth) and trickling filter (fixed film growth) remove many pathogens. The pathogens are adsorbed and entrapped in floc particles in the activated sludge process and biofilm of the trickling filter process. Often, the surface of pathogens is coated with secretions from higher life-forms such as ciliated protozoans, rotifers, and free-living nematodes. These secretions change the surface charge of the pathogens, making them amenable to adsorption to floc particles and biofilm.

Although pathogens are removed from the wastewater in sewer systems, bar screens, comminutors, and grit chambers, the number of pathogens removed by

Wastewater Pathogens, by Michael H. Gerardi and Mel C. Zimmerman
ISBN 0-471-20692-X Copyright © 2005 John Wiley & Sons, Inc.

TABLE 21.1 Fate of Wastewater Pathogens in a Wastewater Treatment System

Consumed by higher life-forms such as rotifers and free-living nematodes
Destroyed by biological, chemical, or physical processes
Inactivated by biological, chemical, or physical processes
Leave the treatment system in bioaerosols, grit, screenings, sludge, or final effluent
Removed by adsorption to inert or biological solids

these conveyance and treatment units is insignificant compared with the number of pathogens that pass through these units. However, cinders and other debris removed from the sewer system as well as screenings and other debris removed from bar screens, comminutors, and grit chambers are contaminated with pathogens. Therefore, appropriate hygiene measures and protective equipment should be used when wastewater personnel are exposed to these units and their associated wastes.

Primary sludge contains a large number and diversity of pathogens. Although removal efficiency for viruses, bacteria, and fungi shows much variation during primary sedimentation, primary clarifier sludge does contain a significant number and diversity of viruses, bacteria, and fungi. Most of these pathogens are removed in primary clarifiers through their adsorption to settleable solids.

Removal efficiencies for protozoan cysts and oocyts and helminth eggs during primary sedimentation are much better than the removal efficiencies for viruses, bacteria, and fungi. Cysts, oocysts, and eggs are more easily removed in primary clarifiers because of their size and density. With increasing detention time in primary clarifiers, cysts, oocysts, and eggs as well as viruses, bacteria, and fungi are removed in increasing numbers. Therefore, primary sludge is heavily laden with pathogens, For example, a dry kilogram of municipal primary sludge may contain over 100,000 viable eggs of *Ascaris lumbricoides*.

Secondary sludge also contains a large number and diversity of pathogens. Secondary sludge contains settled solids in the form of floc particles or biofilm from biological treatment processes. Because of the relatively large surface area of floc particles and biofilm and the secretions of higher life-forms, viruses, bacteria, and fungi are removed easily in large numbers from the wastewater. Therefore, secondary sludge also is heavily laden with pathogens.

Primary and secondary sludges typically are transferred to aerobic and anaerobic digesters for additional treatment. Here, the pathogens in the sludges may be inactivated or destroyed.

INACTIVATION AND DESTRUCTION

Pathogens may be inactivated or destroyed through several biological, chemical, and physical measures. Inactivation is a period of time during which the pathogen is rendered harmless or incapable of causing an infection. If the inactivation period is long enough, the pathogen dies. However, if the inactivation period is not long enough, the pathogen may become active and may cause infection. For example, pathogenic bacteria cannot survive for long periods of time at high pH values (>11) or low pH values (<3). If the retention time of pathogenic bacteria at high or low pH values is short, the pathogens may not be destroyed or inactivated. When the pH value changes to one that permits pathogenic bacteria activation (pH 3–11), the bacteria may reactivate and may become capable of causing infection.

Pathogens within wastewater treatment systems may be inactivated or destroyed by exposure to the following biological, chemical, or physical conditions:

- Anaerobic environment
- Anoxic environment
- Competition for nutrients
- Consumption and digestion by higher life-forms
- Depressed pH (<3)
- Desiccation
- Disinfection
- Elevated pH (>11)
- Elevated temperature
- Entrapment in biological solids
- Heavy metals
- Inability to adapt to a free-living state in the aquatic environment
- Inability to find a suitable host
- Long retention time in treatment units
- Presence of oxidizing agents such as chlorine
- Uv light penetration

An example of inactivation versus destruction of pathogens due to an operational change is the degree of pH increase in lime-treated sludge. An increase in pH to 12.5 rather than 11.5 results in more destruction and inactivation of pathogens and less activation of pathogens when the pH of the sludge drops to near neutral. Also, the longer the pH of the sludge is maintained at 12.5, the greater the number of pathogens destroyed.

Although no definitive statement can be made with respect to the exact degree of pathogen inactivation and destruction in different treatment units, the following statements can be offered:

- Anaerobic digestion of sludge is more effective in inactivating and destroying pathogens than aerobic digestion of sludge.
- Disinfection of the effluent inactivates and destroys many but not all pathogens.
- Increased pathogen inactivation and destruction occur with increasing retention time in treatment units.

Destruction of pathogens via anaerobic digestion of sludge is more efficient than aerobic digestion of sludge—provided short-circuiting of sludge in anaerobic digesters does not occur. The more efficient destruction of pathogens in anaerobic digesters is due to the following operational factors:

- Anaerobic conditions (septicity)
- Elevated temperature

- Higher solids (bacterial) concentration and competition for nutrients
- Increased retention time

In addition to anaerobic digestion of sludge, thermophilic composting of sludge also is an efficient operational method for the destruction of pathogens. Thermophilic composting achieves an increased temperature and a decreased moisture level that destroy a large number and diversity of pathogens. However, these conditions that destroy pathogens also permit the proliferation of *Aspergillus fumigatus*—a fungal pathogen.

Many pathogens survive the adverse biological, chemical, and physical conditions present in a wastewater treatment system. Survival is achieved through short-circuiting (short retention time) in treatment units, complex life cycles of some pathogens, and production of protective features in some pathogens such as capsule development in bacteria and thick shell or hyaline wall production in helminth eggs. Additionally, pathogens consumed but not digested by higher life-forms may survive wastewater treatment systems. These higher life-forms protect the pathogens from adverse conditions and may permit the increase in numbers of pathogens within their bodies.

A wastewater treatment system does not remove, inactivate, or destroy all pathogens that enter the system. A wastewater treatment system should reduce the number of viable pathogens to an acceptable level. This level may be different according to the use of the body of water that receives the wastewater treatment system's effluent or the land that receives the wastewater treatment system's sludge.

22

Disinfection

Disinfection is the destruction of viable, potentially infectious pathogens. Disinfection can be achieved through several operational processes at wastewater treatment systems. Thermophilic processes such as aerobic and anaerobic digesters and sludge composting reduce the number of viable pathogens. Mesophilic aerobic and anaerobic digesters also destroy viable pathogens. However, these processes are used primarily for other purposes such as reduction in sludge volume and destruction of volatile content of sludge, and changes in operational conditions, such as short-circuiting of sludge in and out of digesters, permit the survival of many viable pathogens.

Disinfection of waste streams including the final effluent is achieved with the use of oxidizing agents (Table 22.1) and ultraviolet light (uv). Oxidizing agents include the halogens (chlorine and bromine), hypochlorites of sodium and calcium, chloramines, chlorine dioxide, and ozone. Chlorination/dechlorination is the most commonly used disinfection technology for the disinfection of the final effluent. Chlorine or chlorinated compounds continue to be used frequently for disinfection because of their availability, cost, ease of handling, and ease of application. Ozonation and uv are emerging technologies. Other potentially useful disinfectants include bromine species, iodine (I_2), potassium permanganate ($KMnO_4$), hydrogen peroxide (H_2O_2), metals, and ionizing radiation.

ELEMENTAL CHLORINE

Chlorine is a gas and is an effective alternative for destroying pathogens. Although the use of chlorine produces a variety of chlorinated products including carcinogenic compounds or potential carcinogenic compounds (organochlorine and tri-

---

Wastewater Pathogens, by Michael H. Gerardi and Mel C. Zimmerman
ISBN 0-471-20692-X   Copyright © 2005 John Wiley & Sons, Inc.

TABLE 22.1   Disinfection Alternatives

Alternative	Chemical Formula
Bromine chloride	$BrCl$
Chloramination	NH_2Cl, $NHCl_2$, NCl_3
Chlorine (elemental chlorine)	Cl_2
Chlorine dioxide	ClO_2
Hypochlorination	$NaOCl$, $Ca(OCl)_2$
Potassium permanganate	$KMnO_4$
Ozone	O_3

halomethane), the health risks associated with these products are considered to be less than the health risks associated with the presence of viable pathogens. An additional concern associated with the use of chlorine is the adverse environmental impact of residual chlorine on the receiving water. The adverse impact of residual chlorine is reduced or eliminated by dechlorination with sulfur dioxide (SO_2).

When chlorine demand is satisfied in the final effluent, chlorine reacts with water to form aqueous chorine or hydrochloric acid (HCl) and hypochlorous acid (HOCl) (Eq. 22.1). Hypochlorous acid is the most effective form of chlorine for disinfection. Hypochlorous acid dissociates in water to form the hypochlorite ion (OCl^-) (Eq. 22.2). The hypochlorous acid and hypochlorite ion make up the free chlorine in the chlorine residual test. At pH values >8.5 aqueous chlorine is mostly in the form of the hypochlorite ion (OCl^-). The hypochlorite ion is a less potent disinfectant than hypochlorous acid.

$$Cl_2 + H_2O \text{ ----- } > HCl + HOCl \tag{22.1}$$

$$HOCl < \text{ ----- } > H^+ + OCl^- \tag{22.2}$$

HYPOCHLORINATION

Hypochlorination is the use of sodium hypochlorite ($NaOCl$) or calcium hypochlorite [$Ca(OCl)_2$] for disinfection. Hypochlorite is approximately three times the strength of household bleach. Sodium hypochlorite and calcium hypochlorite are salts of hypochlorous acid and are strong oxidizing agents.

Sodium hypochlorite or liquid bleach is a tinted yellow solution and is available in concentrations between 5 and 15%. Calcium hypochlorite or high-test hypochorite (HTH) is a dry white powder.

CHLORAMINATION

Chloramination occurs when chlorine reacts with ammonia (NH_3) to produce chloramines (Eq. 22.3, 22.4, and 22.5). Chloramines are combined chlorine and have some disinfecting capacity. They are weak disinfectants. Chloramines are the combined available chlorine in the chlorine residual test.

$$NH_3 + HOCl \ \text{-----} > H_2O + NH_2Cl \qquad (22.3)$$

$$NH_2Cl + HOCl \ \text{-----} > H_2O + NHCl_2 \qquad (22.4)$$

$$NHCl_2 + HOCl \ \text{-----} > H_2O + NCl_3 \qquad (22.5)$$

Chloramination is achieved by mixing chlorine (Cl_2) and ammonia (NH_3). There are three chloramines produced through chloramination—monochloramine (NH_2Cl), dichloramine ($NHCl_2$), and trichloramine (NCl_3).

CHLORINE DIOXIDE

Chlorine dioxide (ClO_2) is a gas and an effective disinfectant. It is produced from sodium chlorite ($NaClO_2$) or sodium chlorate ($NaClO_3$). As a disinfectant, chlorine dioxide is more effective than chlorine.

OZONE

Ozone (O_3) is a strong disinfectant and is produced by passing dry air through a system of high-voltage electrodes. Because ozone is unstable, it must be generated on-site. Ozone is highly toxic and reacts quickly to destroy organisms. Ozone inactivates or destroys pathogens by oxidizing the cell wall or genetic material. Ozone also oxidizes structural materials in cell membranes.

ULTRAVIOLET LIGHT

Ultraviolet light or uv radiation destroys pathogens when it is absorbed by cellular proteins, resulting in damage to genetic material. The absorption of uv light results in protein damage and destruction of pathogens. Unlike other disinfectants, uv disinfection does not leave a residual that is toxic to aquatic life.

The most common source of uv light is the low-pressure mercury vapor lamp that emits intense and destructive invisible light that is 254 nm (254 Å) in length. This length or wavelength of light coincides closely with the absorption maximum of nucleic acids, a component of genetic material.

Disinfectants vary with respect to stability and strength (Table 22.2). Regardless of stability and strength, the effectiveness of disinfectants can be diminished by undesired changes in operational conditions (Table 22.3) and pathogen resistance (Table 22.4) to disinfection. Therefore, variations in the stability and strength of disinfectants, undesired changes in operational conditions, and resistance of pathogens to disinfection permit viable pathogens to leave a wastewater treatment system in its final effluent.

Several environmental concerns are related to the disinfection of the final effluent. Chlorinated effluents with poor nutrient removal produce a decreased population of organotrophic bacteria at the point of discharge and a surge in the

TABLE 22.2   Relative Stability and Disinfectant Strength of Disinfectants Based on Coliform Destruction

Stability/Strength	Chemical Disinfectant
Relative stability	Chloramines > chlorine dioxide > free chlorine > ozone
Relative strength	Ozone > chlorine dioxide > free chlorine as HOCl > free chlorine as OCl⁻ > chloramines

TABLE 22.3   Operational Conditions Affecting Efficiency of Oxidizing Disinfectants

Ammonia concentration
Contact time between disinfectant and pathogen
Disinfectant dose or residual
Oxidation demand
pH
Temperature
Turbidity

TABLE 22.4   Trend of General Resistance of Pathogens to Disinfectants

Helminth eggs > protozoan cysts > viruses > bacteria

coliform population downstream of the discharge. This condition is known as "coliform aftergrowth" and occurs as a result of decreased organotrophic bacterial competition because of their sensitivity to chlorine.

Pathogens such as viruses that are entrapped or adsorbed by suspended solids are provided protection from disinfectants by the solids. Pathogens that are associated with cells also are provided protection. Some pathogens in their natural state may encapsulate or clump. The pathogens within the encapsulating material or core of the clumped cells also are provided protection from disinfection.

Some pathogens are more resistant to disinfectants than other pathogens. For example, enteric viruses are more resistant to disinfectants than enteric bacteria. Protozoan cysts are susceptible to disinfection with ozone but are resistant to disinfection with chlorine and uv radiation.

Injured or inactivated pathogens may reactivate through repair mechanisms. Pathogen reactivation or "regrowth" does not appear to be complete, and only a relatively small number of inactivated pathogens do "regrow." Reactivation or "regrowth" of pathogens usually is observed with uv disinfection.

23

Coliform Bacteria and Indicator Organisms

Coliform bacteria (coliforms) are Gram-negative organisms that are normal inhabitants of the intestinal tract of humans and warm-blooded animals. Coliforms may be aerobes or facultative anaerobes that are non-spore-forming, bacillus-shaped bacteria.

Coliforms usually are sparsely concentrated in most habitats (soil and vegetation) except fecal waste. In fecal waste they are highly concentrated. Therefore, the presence of coliforms is considered to be an indicator of fecal contamination.

Coliform bacteria belong in the Family Enterobacteriacae (Table 23.1) and include species in the genera *Citrobacter*, *Enterobacter*, *Escherichia*, and *Klebsiella* (Table 23.2). Perhaps, the most important coliform is *Escherichia coli*. This bacterium is found only in the intestinal tract of humans and warm-blooded animals, and its presence in water historically has been associated with fecal contamination and the possible presence of enteric pathogens. More than 60% of total coliforms are fecal coliforms, and more than 90% of the fecal coliforms are members of the genus *Escherichia*.

The fecal coliform group often is used as an indicator of the potential presence of pathogens. An increasing number or concentration of fecal coliforms provides an indicator of microbial contamination and potential health risk. Besides fecal coliforms, other organisms that serve as indicators of microbial contamination and potential health risk include *Enterococcus*, *Escherichia coli*, *Streptococcus* (fecal streptococci), and total coliforms. These organisms all share good qualities of indicator organisms (Table 23.3). Testing procedures for the use of these organisms as indicators of fecal contamination are provided in the current edition of *Standard Methods for the Examination of Water and Wastewater* (Table 23.4).

Wastewater Pathogens, by Michael H. Gerardi and Mel C. Zimmerman
ISBN 0-471-20692-X   Copyright © 2005 John Wiley & Sons, Inc.

TABLE 23.1   Genera of Bacteria in the Family Enterobactriacae That Contain Pathogens or Potentially Pathogenic Species

Citrobacter
Edwardsiella
Enterobacter
Erwinieae
Escherichia
Hafnia
Klebsielleae
Proteus
Salmonella
Serratia
Shigella
Yersiniae

TABLE 23.2   Coliform Groups

Total Coliforms	Fecal Coliforms
Citrobacter	Citrobacter
Enterobacter	Escherichia
Escherichia	Klebsiella
Klebsiella	

TABLE 23.3   Qualities of Good Indicator Organisms

Associated in large numbers with fecal waste
Easy to count
Easy to identify

TABLE 23.4   Fecal Coliform Tests

Fecal Streptococcus
Membrane filtration
Most probable number (MPN)
Presence/absence

Many coliforms are excellent colonizers because they grow with minimal nutrient requirements and are capable of surviving in harsh environments by encapsulating. Although coliforms are chlorine sensitive, encapsulation protects coliforms from the effects of chlorine. If coliforms are consumed but not digested by higher life-forms such as free-living nematodes, viable coliforms may be found in the excreted wastes of these organisms.

ESCHERICHIA COLI

Escherichia coli is a bacterium in the Family Enterobacteriaceae. Approximately 0.1% of the bacteria in the intestinal tract of humans is represented by Escherichia *coli*. The name Enterobacteriaceae is derived from the Greek *enterikos*—belonging to the intestine. The *Escherichia* comes from the name of Escherich, who first isolated and described the bacterium.

The natural habitat of *E. coli* is the intestinal tract. Together with other bacteria, they produce vitamins that are absorbed by warm-blooded organisms. Vitamins produced by *E. coli* include B and K. However, there are several types of *E. coli* that are opportunistic pathogens and capable of causing disease. These opportunistic pathogens include *E. coli*—enteroinvasive, *E. coli*—enteropathogenic (ETEC), *E. coli*—enterotoxigenic, and *E. coli*—enterohemorrhagic O157:H7.

FECAL STREPTOCOCCI

Although millions of coliform bacteria are found in a liter of raw wastewater, the number of fecal streptococci usually is much less than that of coliform bacteria. Fecal streptococci include enteric bacteria of animals (*Streptococcus avium, S. bovis*, and *S. equinus*) and enteric bacteria of humans (*S. faecalis* and *S. faecium*). These species of streptococci indicate fecal contamination. *S. faecalis* and *S. facecium* indicate fecal contamination of human origin.

Although indicator organisms may be insignificant in number or absent in the final effluent sample of a wastewater treatment plant, this does not guarantee that the effluent is safe. This would only indicate that the effluent was probably safe at the time of sampling. For example, it is difficult to correlate an insignificant number or absence of coliforms with the presence of enteroviruses and protozoan parasites such as *Cryptosporidium* and *Giardia*.

The testing for the presence or absence of indicator organisms is used with much success for the routine evaluation of disinfection of a wastewater treatment plant effluent. Disinfection is not sterilization, and some pathogens do survive a variety of disinfecting processes, including chlorination of the effluent.

Because of the complexity of specific pathogen testing, time delay in obtaining test results, and costs, it is impossible to test for all pathogens that may be present in the final effluent of wastewater treatment plants. Even when specific pathogens are monitored, the absence of these pathogens does not guarantee that other pathogens also are absent. There is no one organism whose absence indicates the absence of all pathogens.

Although current testing of indicator organisms is based on compliance with federal and state standards or permit limits, compliance with these limits may still discharge pathogens into receiving waters. Also, established discharge limits for indicators, for example, 200 MPN (most probable number) fecal coliform/100 ml or 240 MPN total coliform/100 ml, can vary from state to state. Finally, standard testing for indicator organisms may fail to detect dormant pathogens and "nonculturable" pathogens. The increased use of bodies of water that receive effluent from wastewater treatment plants and the increasing pressure for reuse of treated effluent require improvements in the ability to effectively monitor treatment efficiency and the reliability of treatment processes to reduce pathogen numbers.

Part VIII

Hygiene Measures, Protective Equipment, and Immunizations

24

Hygiene Measures and Protective Equipment

There are three distinct stages of pathogen transmission. First, the pathogen must leave an infected individual or host. The pathogen may be found in feces, urine, or blood. Second, the pathogen must be transmitted to a new individual or host. The pathogen may be transmitted to a new individual or host by aerosols, contaminated material, sludge, vector, or wastewater. Third, the pathogen must enter the new individual or host.

Pathogens gain access to a host through a portal of entry. The most common portal of entry is the mucous membrane. There are three common routes of pathogen entrance to a new individual or host. These routes are inhalation (respiratory tract), ingestion (intestinal tract), and invasion (skin and mucous membranes) (Table 24.1). Invasion of the skin may occur through a wound or hair follicle. Entry of pathogens through the mucous membrane (ingestion) is the most common route.

Although the uncontrolled transmission of waterborne diseases has been effectively reduced in the United States, the risk of infection to wastewater personnel is real. Whatever risk may be assigned to any specific pathogenic infection, this risk can be significantly reduced or eliminated through the use of common sense, appropriate hygiene measures, and appropriate protective equipment (Table 24.2). Most hygiene measures are designed to prevent the entrance of pathogens into a new individual.

ANTIMICROBIAL CHEMICAL AGENTS

There are numerous chemical agents that can be used to disinfect hard surfaces such as laboratory counters or act as "antibacterial" soaps or cleaning agents (Table 24.3).

Wastewater Pathogens, by Michael H. Gerardi and Mel C. Zimmerman
ISBN 0-471-20692-X   Copyright © 2005 John Wiley & Sons, Inc.

TABLE 24.1   Examples of Pathogens Associated with Wastewater and Their Routes of Entry to an Individual

Pathogen	Disease	Route of Entry
Clostridium tetani	Tetanus	Invasion
Echovirus	Common cold	Inhalation
Salmonella typhi	Typhoid fever	Ingestion

TABLE 24.2   Measures Available to Wastewater Personnel to Prevent Pathogen Infection

Antimicrobial chemical agents
Automation
Cleanliness and consumption precautions and restrictions
First aid
Proper sampling practices
Protective clothing
Records
Training
Ventilation
Immunization

TABLE 24.3   Antimicrobial Chemical Agents

Chemical Agent	Action(s)
Alcohols	Denature proteins
Alkalis (in soaps)	Denature proteins
Detergents and soaps	Lower surface tension of pathogen; make pathogen susceptible to other agents
Halogens	Oxidize cellular components
Heavy metals (in disinfectants)	Denature proteins
Oxidizing agents	Denature proteins
Phenol and phenolic compounds	Damage cell membranes; denature proteins

These agents lower the surface tension of pathogens, making them susceptible to other destructive agents, denature proteins within the pathogens, or oxidize structural components within the pathogens.

Disinfection of hard surfaces is an important aspect of the control of pathogens in laboratories and lunchrooms. Failure to control pathogens may lead to their transmission. The use of antibacterial soaps or cleaning agents for proper hand washing also is important in the control of pathogens.

The selection of a disinfectant or antibacterial soap is dependent on several factors. These factors include efficacy, cost, ease of use, contact time required, and packaging.

Chemical compounds or antibacterial and anitmicrobial compounds used in soaps behave like antibiotics. These compounds include chlorhexidine, chloroxylenol, ethyl alcohol, gluconate, isopropyl alcohol, and triclosan. Antibiotics attack a specific critical substance within bacteria. The antibacterial agent triclosan, com-

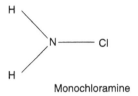

Monochloramine

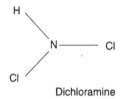

Dichloramine

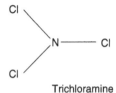

Trichloramine

Figure 24.1 *Chloramines.*

monly used in soaps, works by denaturing an enzyme important in the production of lipids that make up cell walls in bacteria.

Antimicrobial compounds often are conveniently packaged as waterless hand foams, gels, and creams. In these packages the antimicrobial compounds can be taken in the field and used where water is not available for hand washing.

Common disinfectants or general biocides for hard surfaces such as laboratory counters include alcohols, halogens, heavy metals, oxidizing agents, phenolic compounds, and quaternary ammonium compounds. Alcohols used as biocides include 70% ethanol (CH_3CH_2OH) and isopropanol ($CH_3CHOHCH_3$) or rubbing alcohol.

Among the halogen biocides are chlorine and iodine. Chlorine is available as a biocide as sodium hypochlorite (NaOCl) or bleach and calcium hypochlorite [$Ca(OCl)_2$]. Chlorine also is available in the form of chloramines, that is, chlorine combined with ammonia (Fig. 24.1). Iodine is used as a tincture of iodine or iodophors (iodine combined with organic molecules). Heavy metals used in disinfectants because of their antimicrobial ability include copper (Cu), silver (Ag), and zinc (Zn).

Phenol and altered molecules of phenol (phenolic compounds) include hexachlorophene, orthocresol, and orthophenylphenol (Fig. 24.2). Quaternary ammonium compounds or "quats" include benzalkonium chloride and cetyepyridinium chloride (Fig. 24.3). Quats have four organic groups attached to a nitrogen atom. Quats often are mixed with other disinfectants to overcome some problems associated with quats. These problems include loss of effectiveness in the presence of soap and the growth of *Pseudomonas* spp. in the presence of quats.

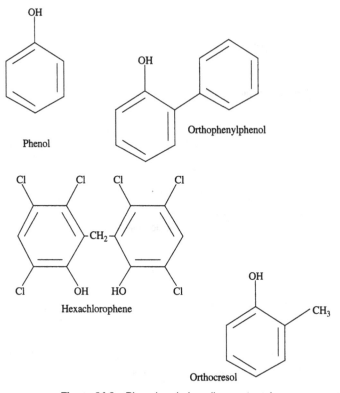

Figure 24.2 *Phenol and phenolic compounds.*

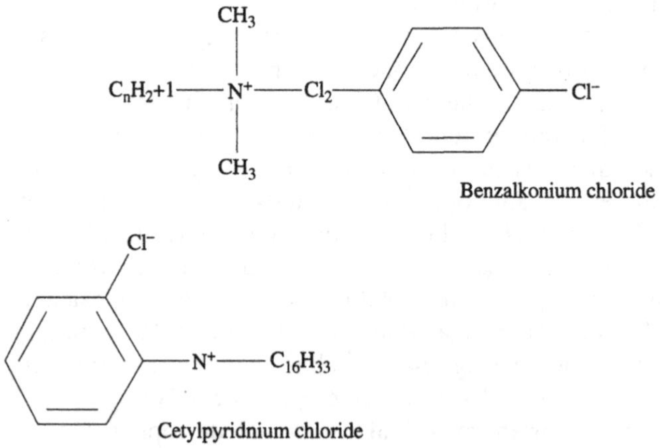

Figure 24.3 *Quaternary ammonium compounds.*

Proper hand washing is an important way to prevent infection. Proper hand washing removes pathogens, removes particles that contain pathogens, and weakens or destroys pathogens. Wastewater personnel should wash their hands before smoking, eating, or touching the mouth or face. Hands should be washed after contact with aerosols, contaminated materials, sludges, and wastewater.

Proper hand washing consists of the use of an antimicrobial soap (surfactant) and warm running water. Cationic and nonionic surfactants are toxic to pathogens, whereas anionic surfactants are less toxic. The wrists, palms, back of hands, and fingers should be washed by rubbing the hands together for at least 20 seconds, being careful not to wash away the soap or foam. Then the hands should be rinsed with warm running water. A fingernail brush should be used to remove debris under the fingernails that may contain pathogens. Wastewater personnel also should shower and shampoo before going home. Hand washing and shower facilities should be provided in locker rooms.

AUTOMATION

Automation, for example, collection of wastewater samples, reduces wastewater personnel contact with wastewaters. The reduction in contact with wastewater also reduces personnel illness rates. The reduction in contact with wastewater can be greatly reduced in areas of high contamination through automation.

CLEANLINESS AND CONSUMPTION PRECAUTIONS AND RESTRICTIONS

There are three important factors that should be exercised to prevent pathogenic infections at wastewater treatment systems. These factors are the use of (1) good common sense, (2) appropriate personal hygiene, and (3) personal protective equipment where appropriate. Personal hygiene and use of personal protective equipment are critical because infections from contact and exposure to pathogens may occur without symptoms.

Because consumption is the most frequent route of pathogenic infections, there are several consumption precautions and restrictions that reflect the use of good common sense. These precautions and restrictions include:

- Avoid touching the ears, eyes, mouth, and nose with your hands, unless you have just washed.
- Confine eating, drinking, smoking, and the use of smokeless tobacco products to designated areas. These areas should be clean.
- Keep your fingernails short; use a stiff soapy brush to clean under your nails.
- Wash your hands frequently and properly after contacting wastewater and before eating, drinking, or smoking, the use of smokeless tobacco products, and at the end of work.
- Wear appropriate gloves where necessary, especially when hands are chapped, cut, or burned.

Gloves should be worn whenever there is contact with wastewater. Often thin flexible, natural rubber latex gloves are used. For some individuals these gloves may cause irritation, allergic reaction, or systemic or anaphylactic reaction. For these individuals, glove liners may be used under the latex gloves and, where practical and appropriate, nonlatex gloves may be used.

Also, no lotion or barrier cream should be used under the latex gloves. The ingredients in the lotion or barrier cream may react with the latex and compromise the integrity of the gloves.

FIRST AID

Immediate first aid should be given to any cut or abrasion that occurs at a wastewater treatment facility. Cuts, abrasions, and burns should receive appropriate first aid. A physician should treat more serious injuries as soon as possible.

PROPER SAMPLING PRACTICES

Proper sampling techniques also help to reduce exposure to pathogens. Sampling techniques should prevent breakage and spillage. Sample bottles should have wide mouth openings and, whenever possible, should be plastic. If glass containers are required, the glass should be coated with plastic. Lids for sampling containers should be tight fitting. Bottles and lids should be cleaned after each use with squirt bottles and paper towels. Carriers for sample bottles should be compartmentalized to prevent breakage. Sampling stations or locations should be hosed down to wash away pathogens that may be present due to spillage or leaking bottles.

PROTECTIVE CLOTHING

Protective clothing (Table 24.4) is effective in preventing infection at wastewater treatment plants. Protective clothing consists of uniforms, shoes or boots, masks, gloves, and goggles. The clothing protects the body when contact with aerosols, contaminated materials, dust, sludges, and wastewater occurs. Protective clothing remains at the wastewater treatment plant and prevents wastewater personnel from bringing pathogens home.

Protective clothing should be washed, dried, and stored at the wastewater treatment plant or clean professionally. High-temperature (57°C) washing machines are needed to kill protozoan cysts. Separate lockers should be provided for work clothes and street clothes. Heavily soiled work clothes should be washed with a copious quantity of bleach.

TABLE 24.4 Protective Clothing

Category	Action or Item
Gloves	Wear appropriate gloves at each work site; elbow-length gloves may be necessary; never submerge top of glove; wash or dispose of gloves after use; wash hands immediately after work when gloves cannot be used
Goggles	Protect eyes from pathogens in aerosols and dust; wash goggles after use
Mask	Prevents inhalation of pathogens in aerosols and dust; ensure proper fit of mask; wash or dispose of mask as directed after use
Uniforms	Leave uniform at work; use separate lockers for work and street clothing; wash work uniforms at work or use professional service; use bleach on heavily soiled uniforms

RECORDS

Safety records should be maintained for all wastewater personnel. Records should contain information addressing accidents, immunizations, major and minor illnesses, and training.

TRAINING

Information about the hazards of pathogens found in wastewater should be provided to wastewater personnel in order for them to be aware that proper hygiene measures and protective equipment are available and should be used to prevent infection. The information should address (1) the pathogens present, especially those from industrial sources such as slaughterhouses, (2) how pathogens are transmitted or enter the body, (3) typical clinical symptoms of gastrointestinal and respiratory diseases, (4) hygiene measures to be used, (5) immunizations available, (6) proper first aid techniques, (7) facility areas that may have the highest risks of pathogen contact, and (8) proper use of protective equipment.

New workers should be screened for proper immunizations. When workers become sick and visit their doctors, the workers should be advised to inform their doctors that they work at a wastewater treatment plant. Also, individuals required to wear latex gloves should receive training with respect to the potential health effects and risks of sensitization related to latex.

VENTILATION

Proper ventilation provides protection from pathogenic agents including allergens and toxins. Pathogenic agents are higher in concentration in poorly ventilated areas compared with outside areas and properly ventilated areas. Appropriate ventilation should be provided for treatment units such as bar screens, grit chambers, lift stations, sludge dewatering facilities, and wet wells.

IMMUNIZATION

The process or procedure by which an individual is provided protection (immunity) from the adverse consequences of invasion or infection of a foreign body, for example, a pathogen, is immunization. Although vaccination is a preventative measure for many pathogens, several of the pathogens, for example, *Leptospira interrogans*, pose decreasing risk over time. Because of the decreasing risks for these pathogens and the adverse effects associated with vaccination and revaccination, some vaccinations may not be recommended. In lieu of vaccination, appropriate personal hygiene measures and appropriate protective equipment are recommended.

25

Immunizations

Immunity refers to the ability of an individual to recognize and defend itself against pathogens. Specific immunity refers to the ability of an individual to defend itself through physiological responses to specific pathogens. The physiological responses are produced through the immune system, which consists of a variety of cells, for example, lymphocytes, and organs, especially the thymus gland.

There are several types of immunity (Table 25.1). Immunity may be genetic (heredity) or acquired (Fig. 25.1). Genetic immunity, also called innate immunity, exists because of genetically determined characteristics. For example, humans do not become infected with many pathogenic agents of animals, and animals do not become infected with many pathogenic agents of humans.

Acquired immunity is obtained in some manner other than by heredity. Acquired immunity may be natural or artificial. Nature immunity may be acquired through exposure to a pathogen (antigen) or the transfer of antibodies to the pathogen (antigen) across the placenta or through the colostrum and breast milk.

Naturally acquired immunity is most often obtained through exposure to a pathogen. Once the pathogen is in the body, the immune system responds to the presence of the pathogen by producing antibodies. The antibodies are found in the immune serum. The immune system also initiates other specific defenses that protect against future infections by the same pathogen (immunity).

Naturally acquired immunity also may be obtained from antibodies transferred to the fetus across the placenta or to an infant in colostrum and breast milk. Colostrum is the first fluid produced by the mammary glands after childbirth.

Artificially acquired immunity is obtained by receiving an injection of the pathogen or immune serum that produces immunity. Artificially acquired immunity may be active or passive. Active immunity occurs when an individual's own immune

---

TABLE 25.1 Types of Immunity

Feature	Type of Immunity		
	Innate	Acquired, Active	Acquired, Passive
Agent(s) providing immunization	Genetics and physiological factors	Antibodies produced by exposure to pathogen	Antibodies produced by another host
Source of antibodies	None	Individual immunized	Another host
Natural	—	Pathogen exposure; antibodies produced or transferred	Serum, breast milk, colostrum
Artificial	—	Vaccination	Injection of gamma globulin or immune serum

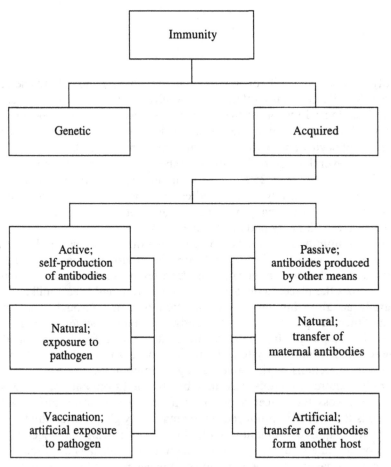

Figure 25.1 *Types of immunity.*

system produces antibodies. Active immunity lasts as long as the antibodies are produced, for example, weeks, months, years, or a lifetime.

Active immunity may be acquired naturally when an individual is exposed to a pathogen. Active immunity also may be acquired artificially when an individual is exposed to a vaccine containing live, weakened, or dead pathogens or the toxins of pathogens. Once exposed to the pathogen (natural or artificial), the individual's immune system responds by producing antibodies. The immune system also "remembers" the pathogen and will respond again if the pathogen enters the body.

Passive immunity is obtained when antibodies are introduced into the body, that is, the body does not produce antibodies, it is not active in the production of antibodies. Passive immunity may be obtained naturally when antibodies are transferred to the fetus across the placenta from the mother's immune system. Passive immunity may be obtained artificially when antibodies produced by other hosts, for example, animals, are introduced to an individual. This is achieved through the injection of gamma globulin or immune serum. Passive immunity usually lasts days to weeks.

Some exotoxins act as antigens. When chemical compounds such as formaldehyde (HCOH) alter the structure of these exotoxins, toxoids are formed. A toxoid is an altered toxin that has lost its ability to cause damage. However, the toxoid still acts as an antigen and can be used to produce immunity when injected into an individual.

Immunoglobulins (Ig) are antibodies. Immunoglobulins bind to pathogens and help pull phagocytes and lymphocytes to pathogens.

Specific immune responses are carried out by lymphocytes. Lymphocytes that are produced in tissue are referred to as bursal equivalent, for example, liver and bone

TABLE 25.2   Immunobiologics Available and Recommended for Special Occupations, Lifestyles, and Travel

Group	Immunobiologics
Health care personnel	Hepatitis B, influenza, polio
Lifestyles (homosexuals, illicit drug users)	Hepatitis B
Travel	Cholera, hepatitis A, measles, meningococcal polysaccharide, plague, polio, rabies, rubella, typhoid, yellow fever

TABLE 25.3   Immunobiologics Commonly Recommended for Wastewater Personnel and Adults

Hepatitis A
Hepatitis B
Influenza (annually for adults 65 years or older)
Measles (for adults born in 1957 or later)
Mumps (if not previously infected)
Pneumococcal (for adults 65 years or older)
Rubella (if not previously vaccinated)
Tetanus and Diptheria (TD) every 10 years after initial dose; after wounds, unless less than 5 years since last dose

marrow are B lymphocytes or B cells. Lymphocytes that are produced in the thymus are T lymphocytes or T cells. Lymphocytes are found in lymphatic tissues (nodules in digestive tract, spleen, lymph nodes, tonsils, adenoids, and thymus) and blood.

Numerous immunobiologics are available for use. The immunobiologics consist of vaccines, immunoglobulins or an immune system booster, and toxoids (Table 25.2).

Immunobiologics recommended or provided for wastewater personnel vary from location to location depending on the incidence of diseases endemic to the location and risk assessment of immunizations and diseases by the local or state health department or care health provider for the wastewater treatment system (Table 25.3). Recommended immunobiologics may include typhoid, paratyphoid, tetanus, and poliomyelitis.

References

American Public Health Association. 1995. *Standard Methods for the Examination of Water and Wastewater, 19th ed.* Washington, DC.

American Water Works Association. 1999. *Waterborne Pathogens*. American Water Works Assoc. Denver.

Berg, G., H. Sanjaghsaz, and S. Wangwongwatana. 1989. Potentiation of the virucidial effectiveness of free chlorine by substances in drinking water. *App. and Environ. Microbiol.* 55.

Benenson, A. S., Ed. 1980. *Control of Communicable Diseases in Man.* American Public Health Association. Washington, DC.

Blacklow, N. R. and G. Cukor. 1982. Norwalk virus: a major cause of epidemic gastroenteritis. *Am. J. Publ. Health.* 72.

Bonadonna, L., R. Briancesco, C. Cataldo, M. Divizia, D. Donia, and A. Pana. 2002. Fate of bacterial indicators, viruses and protozoan parasites in a wastewater multi-component treatment system. *Microbiologica.* 25.

Brakhage, A. A. and K. Langfelder. 2002. Menacing mold: the molecular biology of *Aspergillus fumigatus. Annu. Rev. Microbiol.* 56.

Bridges, B. A. 1976. Survival of bacteria following exposure to ultraviolet and ionizing radiations. In *The Survival of Vegetative Microbes.* T. R. G. Gray and J. R. Posgate, Eds. Cmabridge University Press. Cambridge.

Brinton, M. A. 2002. The molecular biology of West Nile Virus: a new invader of the western hemisphere. *Annu. Rev. Microbiol.* 56.

Britton, G. 1999. *Wastewater Microbiology, 2nd Ed.* Wiley-Liss, Inc. New York.

Buchanan, R. E. and N. E. Gibbons, Eds. 1974. *Bergey's Manual of Determinative Bacteriology, 8th Ed.* The Williams & Wilkins Co. Baltimore.

Burton, N. C. and D. Trout. 1999. *Biosolids Land Application Process, LeSourdsville, Ohio; Health Hazard Evaluation Report No.* 98-0118-2748.

Wastewater Pathogens, by Michael H. Gerardi and Mel C. Zimmerman
ISBN 0-471-20692-X   Copyright © 2005 John Wiley & Sons, Inc.

Calci, K. R., W. Burkhardt III, W. D. Watkins, and S. R. Rippey. 1998. Occurrence of male-specific bacteriophage in feral and domestic animals wastes, human feces, and human-associated wastewaters. *App. Environ. Microbiol.* 64.

Christian C., S. Springthorpe, and S. Sattar. 1999. Fate of *Cryptosporidium* oocyts, *Giardia* cysts, and microbial indicators during wastewater treatment and anaerobic sludge digestion. *Can. J. Microbiol.* 45.

Clancy, J. L., T. M. Hargy, M. M. Marshall, and J. E. Dyksen. 1998. UV light inactivation of *Cryptosporidium* oocysts. *J. AWWA.* 90.

Clark, C. S. 1987. Potential and actual biological related health risks of wastewater industry employment. *J. Water Poll. Control Fed.* 59.

Craun, G. F., Ed. 1986. *Waterborne Disease in the United States.* CRC Press, Inc. Boca Raton.

Dahab, M. F. and R. Y. Surampalli. 2002. Effects of aerobic and anaerobic digestion systems on pathogen and pathogen indicator reduction in municipal sludge. *Water Sci. and Technol.* 46.

De Zuane, J. 1990. *Handbook of Drinking Water Quality; Standards and Controls.* Van Nostrand Reinhold. New York.

Embrey, M. A., R. T. Parkin, and J. M. Balbus. 2002. *Handbook of CCL Microbes in Drinking Water.* American Water Works Association. Denver.

Feachem, R. G., D. J. Bradley, H., H. Garelick, and D. D. Mara. 1983. *Sanitation and Disease; Health Aspects of Excreta and Wastewater Management.* J. Wiley & Sons. New York.

Finstein, M. S. 2004. Protecting watershed from *Cryptosporidium* in manure: a literature review. *Jour. AWWA.* 96.

Gerardi, M. H. 2003. West Nile Virus. *The Keystone Tap.* 2.

Gerardi, M. H. 2002. *Settleability Problems and Loss of Solids in the Activated Sludge Process.* Wiley-Interscience. New York.

Gerardi, M. H., chairman. 1991. *Wastewater Biology: The Microlife.* Water Env. Fed. Alexandria.

Gerardi, M. H. 1982. A review of dangerous gases in sanitary sewer. *Public Works.* 113.

Gerardi, M. H. 1982. Diaphacinone: rodenticide for use in sanitary sewers. *WPCAP Magazine.* 15.

Gerardi, M. H. and M. C. Zimmerman. 1997. Parasites & pathogens. *Operations Forum.* 14.

Grasso, D., chairman. 1996. *Wastewater Disinfection; Manual of Practice FD-10.* Water Env. Fed. Alexandria.

Hickey, J. L. S. and P. C. Reist. 1975. Health significance of airborne microorganisms from wastewater treatment processes; part I: summary of investigation. *J. Water Poll. Control Assoc.* 47.

Hill, D. R. 2001. *Basic Microbiology for Drinking Water Personnel.* American Water Works Association. Denver.

Hoff B. and C. Smith III. 2000. *Mapping Epidemics; A Historical Atlas of Disease.* Grolier Publishing. New York.

James, R. J. and M. Winkler. 1991. *Sludge Parasites and Other Pathogens.* Ellis Horwood Ltd. West Sussex.

Johnston, H. J. 1998. Composting marketing in New York State. *Clearwaters.* 28.

Karpiscak, M. M., L. R. Sanchez, R. J. Freitas, and C. P. Gerba. 2001. Removal of bacterial indicators and pathogens from dairy wastewater by multi-component treatment system. *Water Sci. and Technol.* 44.

Kiple, K. F., Ed. 1993. The Cambridge World History of Human Disease. Cambridge Univeristy Press. Cambridge.

Kowai, N. E. 1982. *Health Effects of Land Treatment: Microbiological.* EPA-600/1-82-007. US EPA. Cincinnati.

Lang, M. 1998. Summary of federal biosolids management regulations. *Clearwaters.* 28.

Lue-Hing, C. 1999. *HIV in Wastewater: Presence, Survivability, and Risk to Wastewater Treatment Plant Workers.* Water Env. Fed., Alexandria.

Manahan, S. E. 1994. *Environmental Chemistry, 6th Ed.* CRC Press LCC, Boca Raton.

Mara, D. and S. Cairncross. 1989. *Guidelines for the Safe Use of Wastewater and Excreta in Agriculture and Aquaculture.* World Health Organization. Geneva.

Oropaza, M. R., N. Cabirol, S. Ortega, L. P. C. Ortiz, and A. Noyola. 2001. Removal of fecal indicator organisms and parasites (fecal coliforms and helminth eggs) from municipal biologic sludge by anaerobic mesophillic and thermophilic digestion. *Water Sci. and Technol.* 44.

Orsini, M., P. Laurenti, F. Boninti, D. Arzani, A. Ianni, and V. Romano-Spica. 2002. A molecular typing approach for evaluating bioaerosol exposure in wastewater treatment plant workers. *Water Research.* 36.

Payment P., R. Plante, and P. Cejka. 2001. Removal of indicator bacteria, human enteric viruses, *Giardia* cysts, and *Cryptosporidium* oocysts at a large wastewater primary treatmnet facility. *Can. J. Microbiol.* 47.

Regli, S., J. B. Rose, C. N. Haas, and C. P. Gerba. 1991. Modeling the risk from *Giardia* and viruses in drinking water. *Jour. AWWA.* 83.

Sack, R. B. 1975. Human diarrheal disease caused by enteropathogenic *Escherichia coli. Ann. Rev. Microbiol.* 29.

Spellman, F. R. 1999. *Choosing Disinfection Alternatives for Water/Wastewater Treatment.* Technomic Publishing Co. Lancaster.

Spellman, F. R. 1997. *Microbiology for Water/Wastewater Operators.* Technomic Publishing Co. Lancaster.

Schmidt, G. D. and L. S. Roberts. 2000. *Foundations of Parasitology, 6th Ed.* McGraw-Hill. Boston.

Sheppard, H. W. and M. S. Ascher. 1992. The natural history and pathogenesis of HIV infection. *Annu. Rev. Microbiol.* 46.

Silver, S. 1996. Bacterial heavy metal resistance: new surprises. *Annu. Rev. Microbiol.* 50.

Smith, E. H. and R. C. Whitman. 1992. *NPCA Field Guide to Structural Pests.* National Pest Control Association. Dunn Loring, Virginia.

Sobsey, M. D., F. Takashi, and R. M. Hall. 1991. Inactivation of cell-associated and dispersed hepatitis A virus in water. *Jour. AWWA.* 83.

Spillman, S. K., F. Traub, M. Schwyzer, and R. Wyler. 1987. Inactivation of animal viruses during sewage sludge treatment. *App. and Env. Microbiol.* 53.

Tanner, R. S. 1989. Comparataive testing and evaluation of hard-surface disinfectants. *Jour. Ind. Microbiol.* 4.

Teltsch B. and E. Katzenelson. 1978. Airborne enteric bacteria and viruses from spray irrigation with wastewater. *App. Environ. Microbiol.* 35.

Theerman, J., chairman. 1991. *Biological Hazards at Wastewater Treatment Facilities.* Special Publication of Task Force on Biological Hazards. Water Environment Federation. Alexandria.

Thorn, J. L. Beijer, T. Johsson, and R. Rylander. 2002. Measurement strategies for the determination of airborne bacterial endotoxin in sewage treatment plants. *Ann. Occp. Hyg.* 46.

Trout, D., C. Muller, L. Venczei, and A. Krake. 2000. Evaluation of occupational transmission of hepatitis A virus among wastewater workers. *Jour. Of Occupational and Env. Medicine.* 42.

Venczel, L., S. Brown, H. Frumklin, J. Simmonds-Diaz, S. Deithchman, and B. Bell. 2003. Prevalence of hepatitis A virus infection among sewage workers in Georgia. *Am. Jour. Inds. Medicine.* 43.

Wallis, P. M. and D. L. Lehmann. 1983. *Biological Health Risks of Sludge Disposal to Land in Cold Climates.* University of Calgary Press. Alberta.

Zaman V. and L. A. Keong. 1982. *Handbook of Medical Parasitology.* ADIS Health Science Press. Balgowlah, Australia.

Abbreviations and Acronyms

AIDS	Acquired immunodeficiency syndrome
°C	Degrees Celsius
CDC	Centers for Disease Control and Prevention
CFR	Code of Federal Regulations
DEET	*N,N*-diethyl-*m*-toluamide
DNA	Deoxyribonucleic acid
HTH	High-test hypochlorite
I/I	Inflow and infiltration
MPN	Most probable number
MRSA	Methicillin-resistant *Staphylococcus aureus*
mm	Millimeter
nm	Nanometer
RNA	Ribonucleic acid
sp.	Species (one)
spp.	Species (two or more)
μm	Micron
uv	Ultraviolet light
U.S. EPA	United States Environmental Protection Agency
USDA	United States Department of Agriculture
VOC	Volatile organic compounds
WHO	World Health Organization

Wastewater Pathogens, by Michael H. Gerardi and Mel C. Zimmerman
ISBN 0-471-20692-X   Copyright © 2005 John Wiley & Sons, Inc.

Chemical Compounds and Elements

Ag	Silver
BrCl	Bromine chloride
Ca^{2+}	Calcium ion
$Ca(OCl)_2$	Calcium hypochlorite
CH_4	Methane
CH_3CH_2OH	Ethanol
$CH_3CHOHCH_3$	Isopropanol
$CH_3CHOHCOOH$	Lactic acid
CH_3COOH	Acetic acid
$C_8H_{13}N$	Indole
C_9H_9N	Skatole
$CH_3NCH_3CH_3$	Trimethylamine
CH_3SH	Methyl mercaptan
C_2H_5SH	Ethyl mercaptan
Cl_2	Chlorine
ClO_2	Chlorine dioxide
CO	Carbon monoxide
CO_2	Carbon dioxide
Cu	Copper
H^+	Hydrogen ion or proton
HCl	Hydrochloric acid
HCOH	Formaldehyde
HCOOH	Formic acid
$H_2NCH_2NH_2$	Cadaverine
$H_2N(CH_2)_4NH_2$	Putresine

Wastewater Pathogens, by Michael H. Gerardi and Mel C. Zimmerman
ISBN 0-471-20692-X   Copyright © 2005 John Wiley & Sons, Inc.

H_2O	Water
H_2O_2	Hydrogen peroxide
HOCl	Hypochlorous acid
H_2S	Hydrogen sulfide
I_2	Iodine
$KMnO_4$	Potassium permanganate
NaOCl	Sodium hypochlorite
NCl_3	Trichloramine
$NHCl_2$	Dichloramine
NH_2Cl	Monochloramine
NH_3	Ammonia
NO_3^-	Nitrate ion
O_2	Oxygen
O_3	Ozone
OCl^-	Hypochlorite ion
P	Phosphorus
SO_2	Sulfur dioxide
SO_4^{2-}	Sulfate ion
Zn	Zinc

Glossary

acute Having rapid onset, severe symptoms, and a short course

aerobic Living in the presence of oxygen

agranulocyte A nongranular leukocyte

alkali A substance that can neutralize an acid

anaerobic Living in the absence of oxygen

anionic Having a net negative charge

anorexia Seen in malaise, commencement of all fevers and illnesses, also in disorders of the digestive tract, especially the stomach

anoxic Living in the presence of nitrate

antibiotic Any substance, such as penicillin, produced by or derived from a microorganism with the ability to inhibit or kill another microorganism

antibody Any protein in the blood that is generated in reaction to a foreign protein or carbohydrate, neutralizes it, and thus produces immunity against certain microorganisms or their toxins

anticoagulant Preventing the process of clotting

antigen Any substance that, on being introduced into the body, stimulates the production of antibodies

arthropod An animal that has a hard, jointed exoskeleton and paired, jointed legs

asymptomatic Without disease symptoms

autotrophic An organism that obtains carbon from carbon dioxide

avian Bird

bacillus A rod-shaped bacterium

bacteriophage A virus whose host is a bacterium

Wastewater Pathogens, by Michael H. Gerardi and Mel C. Zimmerman
ISBN 0-471-20692-X Copyright © 2005 John Wiley & Sons, Inc.

bile A thick, viscous liquid secreted by the liver

bioaerosol An aerosol that contains a viable pathogen

bionomen The scientific name of an organism, for example, *Rattus rattus*

brucellosis A widespread, infectious disease of cattle, goats, and pigs that is caused by bacteria in the genus *Brucella*. Referred to as undulant fever or Malta fever in humans

capsid A protective lipid covering on the outside of some viruses

capsule A protective structure that surrounds the cell. It is almost always composed of polysaccharides. The capsule protects the bacterium form phagocytic white blood cells

carrying capacity The number of organisms that an area or volume of the environment can support at any given time

cationic Having a net positive charge

cerumen The waxlike, soft brown secretion found in the external canal of the ear

cirrhosis A chronic disease of the liver

coccus A spherical-shaped bacterium

colic Spasm in any hollow or tubular soft organ

colostrum Secretion from the breast before the onset of true lactation 2 or 3 days before delivery

commensal One of two organisms that live together in a nonparasitic state

conidium An asexual spore of fungi

cosmopolitan Worldwide in distribution or occurrence

cyst A thick, resistant covering secreted by protozoans for protection against harsh environmental conditions

cysticercus The encysted larval form of a tapeworm consisting of a rounded cyst or bladder into which the scolex is invaginated

cytoplasm The jellylike contents of a cell that is surrounded by the cell membrane

dermatitis Inflammation of the skin

desiccation Drying out

dysentery A term applied to a number of intestinal disorders, especially of the colon, characterized by an inflammation of the mucous membrane

ectoparasite A parasite that lives on (outside) its host

encephalitis Inflammation of the brain

endemic Consistently present in a region

endoparasite A parasite that lives in (inside) its host

endospore A thick-walled spore within a bacterium

enteric To the intestinal tract

envelope A protective lipid covering on the outside of some viruses

enzyme Any of several complex proteins that are produced by cells

epidemic An infectious disease that attacks many people at the same time in the same area

epithelium Cells that form the outer surface of the body and line the body cavities and principal tubes and passageways leading to the exterior of the body

erythrocyte   Red blood cell or corpuscle

etiology   Study of the causes of disease that result from an abnormal state producing pathological conditions

facultative anaerobe   An organism with the ability to live with or without oxygen

flora   Population of the organisms within an area

gangrene   The putrefaction of soft tissue; a form of necrosis

gastoenteritis   Inflammation of the stomach and intestinal tract

globulin   One of a group of simple proteins insoluble in pure water but soluble in neutral solutions of salts or strong acids with strong bases

granulocyte   A granular leukocyte

halogen   A salt former; one of a group of elements (bromine, chlorine, fluorine, and iodine) having similar chemical properties

helminth   Worm

heparin   An acid produced by the liver and some leukocytes that inhibits coagulation

hepatitis   Inflammation of the liver by virus or toxic origin

histamine   A substance found in the body wherever tissues are damaged

hyaline wall   A thick, resistant covering surrounding the cyst or eggs of some parasitic protozoans and helminths

incubation period   Interval between exposure to infection and appearance of the first symptoms

indigenous   Native to a region

infective dose   Number of pathogens required to initiate infection of a host

jaundice   A condition characterized by yellowness of skin, whites of eyes, mucous membranes, and bodily fluids due to deposition of bile pigment

larva   Applied to a young animal, such as a worm or insect, that differs in form from the parent

lavage   Washing out of a cavity

lesion   An injury or wound

lumen   The space within an artery, vein, intestine, or tube

lysis   Splitting or breaking apart

lysozyme   A substance present in tears, saliva, and other bodily fluids that has antibacterial activity

macrophage   A cell having the ability to phagocytose particulate substances

malignance   Severe form of occurrence, tending to grow worse

masseter   The muscle that closes the mouth

mast cells   Connective tissue cells that contain heparin and histamine in their granules

meningitis   Inflammation of the membranes of the spinal cord or brain

mucin   A glycoprotein found in mucus

mutation   An alteration of the genes or chromosomes of an organism or nucleic acid material in a virus

narcosis   An unconscious state

nematode A roundworm

nonionic Having no net charge

obligate Required

olfactory Pertaining to smell

omnivorous Living on all kinds of food

oncosphere Embryonic stage of a tapeworm that has hooks

oocyst The encysted form of a fertilized gamete that occurs in some protozoans

organic Molecules that contain carbon and hydrogen

organochlorine Molecules that contain carbon, hydrogen, and chlorine

ovum Egg

pandemic A disease that occurs over a large geographic area, such as a continent, and affects a high percentage of the population

pasteurization A process of sterilization that exposes liquids or solids to specific high temperature for a period of time in order to destroy certain organisms

pathogen An infectious agent that causes disease

parasite An organism that lives on or in another organism and obtains its food and shelter from that organism

pathogenesis Origination and development of a disease

pediculosis An infection of lice

peristalsis A progressive wavelike movement that occurs involuntarily in the hallow tubes of the body

phagocytosis Ingestion and destruction of particulate material such as bacteria and protozoans

physiological Bodily functions

platelet A round or oval disk, approximately 1/2 the size of an erythrocyte, found in the blood

pulmonary Involving the lungs

proglottid A segment of a tapeworm

prothrombin A chemical substance in circulating blood that is used to produce thrombin that is used in forming blood clots

saprophytic Living or growing in decaying or dead matter

sclera The white outer coating of the eye

scolex The head of a tapeworm

scutum Plate of bone resembling a shield

sebaceous Having an oily, fatty nature

sebum A fatty secretion of the sebaceous glands of the skin

simian A primate

spirillium A spiral-shaped bacterium

spore A specialized, resistant resting cell produced by a bacterium or fungus

stupor A condition of unconsciousness, torpor, or lethargy with suppression of sense or feeling

subclinical Asymptomatic

synovia   The joint-lubricating fluid

toxoid   A toxin treated so as to destroy its toxicity, but still capable of inducing the formation of antibodies on injection

trophozoite   A protozoan nourished by its host during its growth stage

tropism   Turning toward

vector   An organism, such as an insect or rodent, that transmits a pathogen

viremia   Virus in the blood

virulent   Able to overcome a host's defensive mechanisms

viscera   Internal organs

zoogloeal   Rapid growth of floc-forming bacteria in activated sludge and trickling filter processes

zoonotic   Relating to disease transmission from animals to humans

Index

Wastewater Pathogens, by Michael H. Gerardi and Mel C. Zimmerman
ISBN 0-471-20692-X   Copyright © 2005 John Wiley & Sons, Inc.

Printed in the United States
By Bookmasters